Microelectronic Systems F1

The other titles in the BTEC Microelectronics Series are:

MICROELECTRONICS NII by D. Turner

MICROELECTRONICS NIII by D. Turner

PRACTICAL EXERCISES IN MICROELECTRONICS by D. Turner

MICROPROCESSOR INTERFACING by G. Dixey

MICROCOMPUTER SYSTEMS by R. Seals

MICROPROCESSOR-BASED SYSTEMS by R. Seals

Microelectronic Systems F1

Geoff Cornell BSc, DipEdMan, CEng, FIEE, CertEd

Stanley Thornes (Publishers) Ltd

First published in 1992 by:
Stanley Thornes (Publishers) Ltd
Old Station Drive
Leckhampton
CHELTENHAM GL53 0DN
England

British Library Cataloguing in Publication Data

Cornell, Geoff
Microelectronic Systems
I. Title
621.381

ISBN 0-7487-1281-X

Typeset by Florencetype Ltd, Kewstoke, Avon.
Printed and bound in Great Britain at The Bath Press, Avon.

Contents

Preface

This book is aimed at those students who are working towards one of the First Awards in Engineering of the Business and Technician Education Council (BTEC). As many young people are being introduced to BTEC units and programmes earlier and earlier it is hoped that this book would also be of use in schools and colleges delivering TVEI, CPVE* and GCSE and other schemes where vocational relevance is both demanded and built into the curriculum.

A knowledge of the importance of microelectronic systems to modern industry is essential. BTEC thinking is to challenge and motivate the students on their courses by encouraging a dynamic, problem solving approach based on practical assignments. Students will find encouragement in developing their own ideas about the role of microelectronics, not only in the systems they may meet today but also in the as yet unknown applications of the future.

It is intended to cover the content of the standard BTEC devised half-unit at first level, number 2869B, Microelectronic Systems and also to provide a small overlap with some of the material from the national level units in microelectronic systems.

The pace of microelectronic development is increasing daily. The technology currently supports a range of microprocessors of up to 32 bits in length. The content of this book will be restricted to 8-bit processors, in particular the Motorola 6802 and the Intel 8085. Acknowledgement is given to both companies for their permission in printing extracts from their data sheets and instruction sets.

The author has many years of experience of teaching and helping students to learn about modern microelectronic systems, having wide knowledge of the further education system, particularly in delivering BTEC courses at all levels. He has had first hand experience of curriculum development and program validation for BTEC. He has helped many students to learn through their own experiences in the work place and has successfully delivered many courses in that environment.

Geoff Cornell Woodley

* TVEI is the Technical and Vocational Education Initiative, designed as a 14–18 age group provision between schools and colleges. CPVE is the Certificate for Pre-Vocational Education intended for the 14–16 age group. Both of these schemes were intended to make the experiences of young people in schools more realistic and relevant to the needs of those industries that would eventually be employing them.

Acknowledgements

This book could not have been written without a great deal of help, advice and encouragement from my family, friends and colleagues.

To Andy Thomas and Brian Carvell, who originally commissioned the book, my thanks for giving me the opportunity to do something I have always wanted to do and keeping me out of mischief at the same time!

To Intel and Motorola for allowing me to use extracts from their data sheets and other publications to complement the text, I extend my grateful thanks.

To my many colleagues who, over the years, have given me of their time and knowledge to help me develop my own experience in microelectronics, I shall always be grateful.

To the very many students who, unknowingly, have helped me to forge that knowledge into a real experience over the past years, a large debt of gratitude is owed.

Finally, the text could never have seen daylight if it had not been for the consistent support and persuasion of my family and particularly the unfailing support of my wife who has typed the manuscript.

ONE

Systems

OBJECTIVES

At the end of this unit all students should be able to:

Diagnose examples of processing inputs and determine the operational functions of:

input systems
processing units
output systems
transducers

Systems form part of our everyday lives. There are large and small ones, simple and complex ones, cheap and expensive ones. Examples of systems include cassette tape recorders, personal stereo cassette players, compact disc players, microcomputers, domestic appliances of various types, active suspension units for racing cars, military weapons, communications and satellites, spacecraft and large industrial complexes such as oil refineries.

There are many ways of explaining what a system is or what it contains. All systems, no matter how large or small require *energy* in some form or other. A system uses *energy* supplied from somewhere outside itself to *process*, or change, any *input* which is applied to it. The result is an *output* from the system, which is a changed, or modified, form of the input. Later on in this chapter you will meet some familiar things represented as systems.

In Figure 1.1 opposite is shown a block diagram. This is used as a way of simplifying something, in this case the idea of energy and process. It is done by placing labels in rectangular shapes called *blocks*. This method will be used many times throughout the book.

This simplified block diagram is used to show the basic components of any system, i.e. the inputs, the energy supplied, the process involved that uses the energy to act on the inputs and finally the results of the process called outputs. All systems, no matter how large or small, may be drawn in this way.

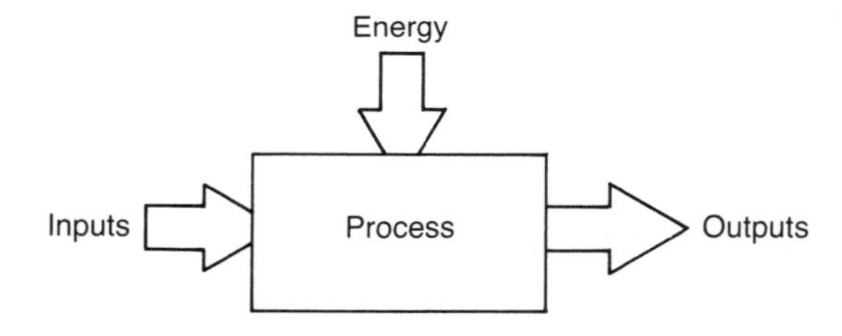

Figure 1.1 Block diagram representation of a system

Now, a word of caution. Later on in the book, in Chapter Two, the idea of *flow diagrams* will be met. These are used by engineers and technicians to simplify (reduce or break down into smaller parts called 'sub-systems') a more complex system. They are also used by computer technicians to help them write and interpret computer programs where they want to see each step or stage of a program developed in a logical way. It will be obvious that the rectangular shape used to show *process* in a system block diagram is similar to that used in flow diagrams. There is no need for concern here; there is no confusion as each uses this shape to show a process, i.e. something happening. Also the situation in which it is used will make it clear whether a diagram is a block diagram or flow chart.

The inputs can be of any number, as can be the outputs. There is also no need for the number of inputs to equal the number of outputs.

The processes involved can be of many different types. They could be electronic, electrical, mechanical, hydraulic, pneumatic, chemical, biological, nuclear, etc. Any type of process that can be thought of is probably used in some system somewhere or other.

Examples of typical systems may now be considered to see if they can be made to fit the simplified model shown in Figure 1.1.

(c) amplify the small electrical signal. (*Amplifier*)
(d) provide further amplification so that the electrical signals are in a suitable condition to feed the loudspeakers. (*Power amplifier*)

The outputs are the usual two loudspeakers and/or earphones for the stereo signal. These are other examples of *transducers* which will be seen more and more as a very important part or component to make any system work properly.

A transducer is simply a device which turns one form of signal into another form, i.e. magnetic to electrical, electrical to sound in the example of the cassette tape recorder.

CASSETTE TAPE PLAYER

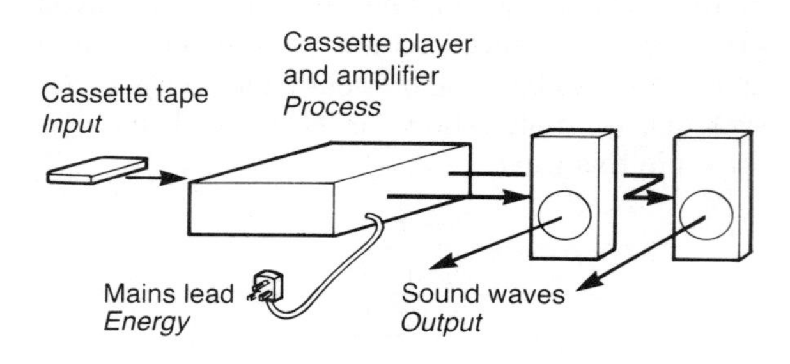

Figure 1.2 Cassette taper player

The energy supply is shown from the a.c. domestic mains, but could equally well be from batteries, provided that the cassette player and amplifier have a suitable means of changing between a.c. and d.c. so that the internal components will not be damaged.

The input is a signal derived from the magnetic tape in the cassette itself. This will have a signal already recorded on it which could be in the form of speech or music.

The cassette player and amplifier have to do a number of things which include the following:

(a) make the tape move from one reel to another. (*Motor*)
(b) pass the tape over the playback head which converts the magnetic signal to an electrical one. (*Transducer*)

MICROWAVE OVEN

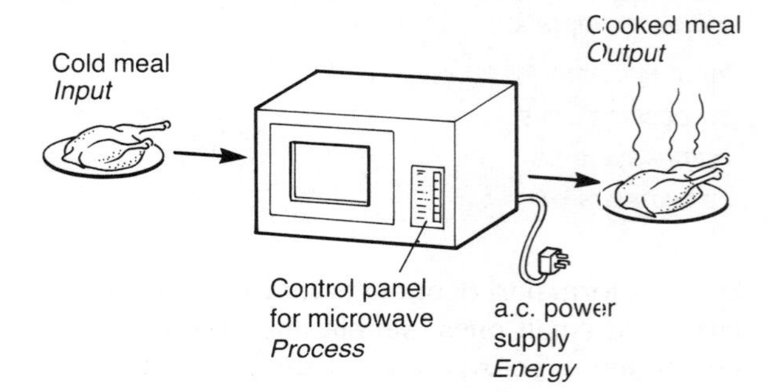

Figure 1.3 Microwave oven

The energy supply comes from the a.c. domestic mains. The input could be a cold, precooked meal, frozen food, which needs defrosting or a meal or item which needs cooking; a number of possible input situations could exist. The output in each case is exactly the same item but hopefully hot, defrosted or cooked. In this case there has been no need of any transducers, the inputs and outputs being the same things.

The process inside the microwave is quite complex and the user has to *control* the process by deciding *how long* the process should take and the *power level* the process is to be operated at.

The control panel can be a large knob or touch sensitive panel but most modern microwaves control the power of the microwave generator and the length of cooking time in some way or other.

DOMESTIC WASHING MACHINE

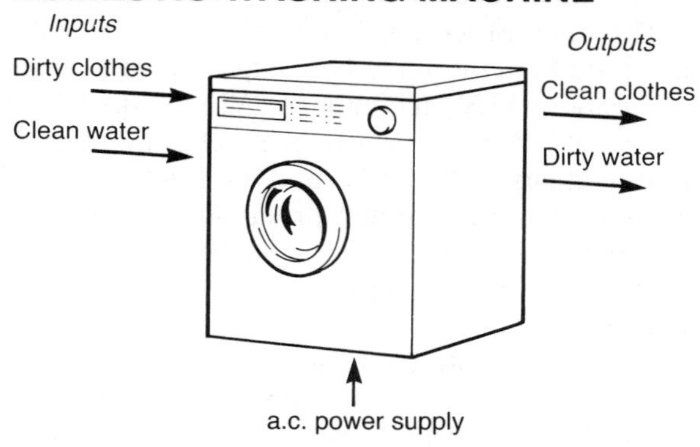

Figure 1.4 Domestic washing machine

In this example, the process, or processes, which take place can be quite complicated and should include the following jobs:

(1) Fill with water – this needs a *level detector*; an *on/off controller*.
(2) Heat the water – this needs a *heater*; a *thermostat*.
(3) Rotate the drum – this needs a speed control; an *on/off controller*.
(4) Drain waste water – this needs a *level detector*; an *on/off controller*.
(5) Clock control is needed to *time* the various sequences.

A number of additional factors have to be considered, in particular the *safety* of the system and the *user*. A number of *transducers* and *controllers* are needed to make sure the whole system and its sub-systems work efficiently and safely.

The inputs and outputs are, once again, the same items but different in that the clothes are clean and fresh.

In the above examples, it has been seen that the inputs and outputs are the same items, but changed in some way. Also, the amount, type and variety of processing have been seen to become more and more complex, depending upon what the system is called on to do.

Each system has its own clearly seen inputs and outputs. Each one needs a clearly identifiable energy source to do the job required of the system. Each one poses its own hazards and threats to personal safety, which must be dealt with before users can trust themselves to use the system.

All of the above systems have used electrical and/or electronic methods of controlling the processes in the systems. Although there are many systems which use other types of transducers and other means of control the trend is increasingly towards electronic operation and control. Electronics is a very wide term and is taken to include modern microelectronics.

In the modern world in which we live and work it has become increasingly obvious that electronics is taking over more and more operations and functions previously carried out by other means. There are, however, still many things that, as yet, electronics cannot do. Applications which need large amounts of power to operate them will often use hydraulic or pneumatic methods. Large electric motors are still needed to drive large loads such as radar aerials, etc.

Where electronics has taken a large step forward is in the application of microelectronics to the control of systems in general. The microprocessor in particular has brought processing power to many situations that previously needed large mainframe computer facilities to control the processes involved.

ANALOGUE AND DIGITAL SYSTEMS

Up to this point no mention has been made of how these processes are to be controlled. A number of different ways could be considered and each could be carried out using either *analogue* or *digital* electronics or, in a real situation, a mixture of both. You may be familiar with these terms, but it is useful to know some of the differences between the two.

The real world is an analogue one. The signals that occur there are usually continuous and gradually changing. An example is that of an audio signal shown in Figure 1.5 below.

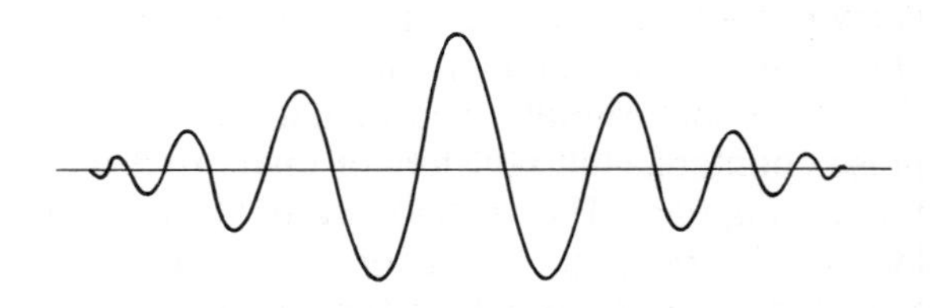

Figure 1.5 Analogue signal waveform

The analogue signal waveform shown in Figure 1.5 is continuous (it has no breaks in it!) and is varying above and below a straight line. This is how such a signal would appear on the screen of a cathode ray oscilloscope (CRO). This is a very useful instrument used for displaying and measuring very rapidly changing electrical signals.

The horizontal line represents *earth potential*, or *ground* as some people call it. It has a value of 0 V. Any part of the signal above this line has a positive value at that instant and any part below the line has a negative value. This idea will be developed more in Chapter Three.

A digital system is one which will allow only one of two possible states. Consider the pulse train shown in Figure 1.6 below.

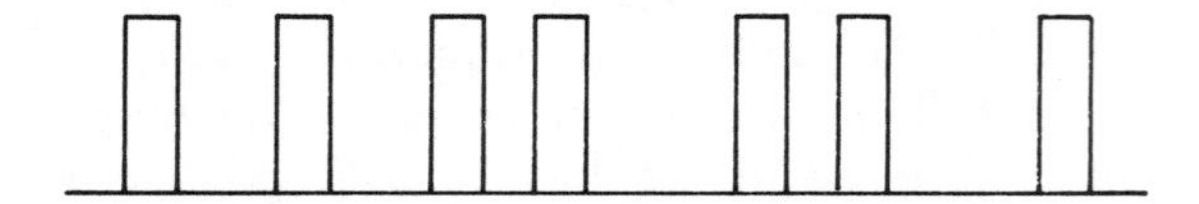

Figure 1.6 Digital signal waveform

Figure 1.6 shows a series of pulses, called a *pulse train*. These can also be shown on a cathode ray oscilloscope screen. This signal is definitely *not* continuous because it has breaks, or gaps, in its waveform. Also, the pulses exist above the line, so when they are present, they are positive in value. They are obviously zero when they are *not* present. There are no possible states between these two in a digital waveform.

A number of integrated circuits are available in both analogue and digital forms. These integrated circuits are sometimes called *microchips* or *chips* for short. They consist of lots of components made very small on a very small area of material and covered with a protective coating. They can carry out a large range of applications. Analogue chips can act as amplifiers, filters or signal level detectors, for example, and digital chips can be pulse counters, shift registers or memory devices for a computer. The technology of their manufacture is beyond the scope of this book. Knowledge of these integrated circuits is best gained by using them in real situations where they will control events or solve problems for us.

The microprocessor is an example of a digital integrated circuit made by using very large scale integration (VLSI) techniques. (VLSI is described on p. 38.) It processes in parallel a number of binary digits called *bits* and in fractions of a second. It is the parallel processing power that has made the microprocessor such a versatile tool to technicians and engineers in a number of industries.

The major feature of the microprocessor that has made it supreme when compared with other integrated circuits of the past is that it is *programmable*. This allows the user the flexibility to make it carry out so many functions and be adaptable if needed. Once a solution to a problem is found in an engineering situation, it usually follows that improvements are found necessary. This usually means a complete rethink but a microprocessor may be re-programmed to take on board a revised solution.

CONTROL OF CENTRAL HEATING

Modern houses pose a number of problems for the designer of central heating systems. These problems are due to the design of the structure of the house, the way it faces, where it is situated and most importantly, the needs of the occupants.

Conventional thinking accepts a temperature of 22 °C in the living rooms, 18 °C in the halls and landing areas and 15 °C in the bedrooms. This is not to everyone's taste and is difficult to achieve in practice.

Coupled with this is the problem of control of the temperature of the hot water from the taps in the kitchen, cloakroom and bathroom. The risk of scalding must be eliminated. This indicates that a difference is needed between the temperatures of the domestic hot water and the water circulating round the heating system.

Some form of localised control could well be used. A room thermostat could control a valve in each radiator to open or close it when a change from the desired value is sensed. This is often not sufficiently sensitive for some users.

A microprocessor based controller could be designed and programmed to read a number of temperature sensors both inside and outside the house and to provide controlling action accordingly. Complete control over not just 24-hour periods but complete seasonal cycles is possible, usually resulting in a more efficient use of fuel.

A simplified diagram is show in Figure 1.7 below.

A microprocessor is an example of a *digital programmable* integrated circuit. It needs transducers which will convert the analogue signals received by the sensors into a digital form it can recognise, read, store and process. Usually some form of analogue-to-digital conversion is built into the interface device between the sensors and the microprocessor.

Once the microprocessor has carried out its control function, signals need to be sent to various actuating devices, such as the pump, control valves and thermostats. These devices do not normally recognise digital signals, so some form of digital-to-analogue conversion process is needed in the output interface to perform that function.

Summary

It has been suggested that many modern domestic devices are systems in their own right. All have certain features which can be described as follows overleaf.

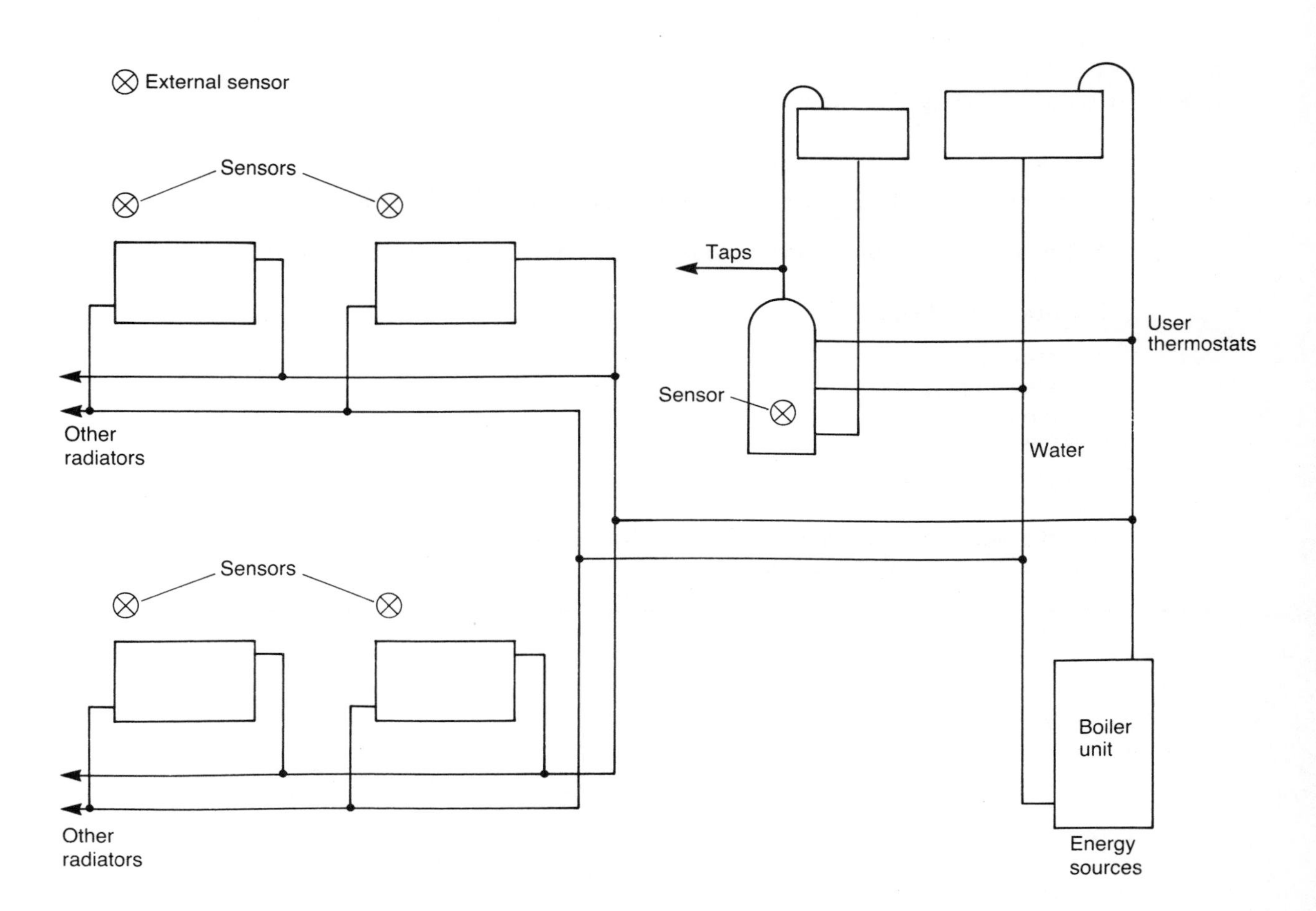

Figure 1.7 Domestic central heating system

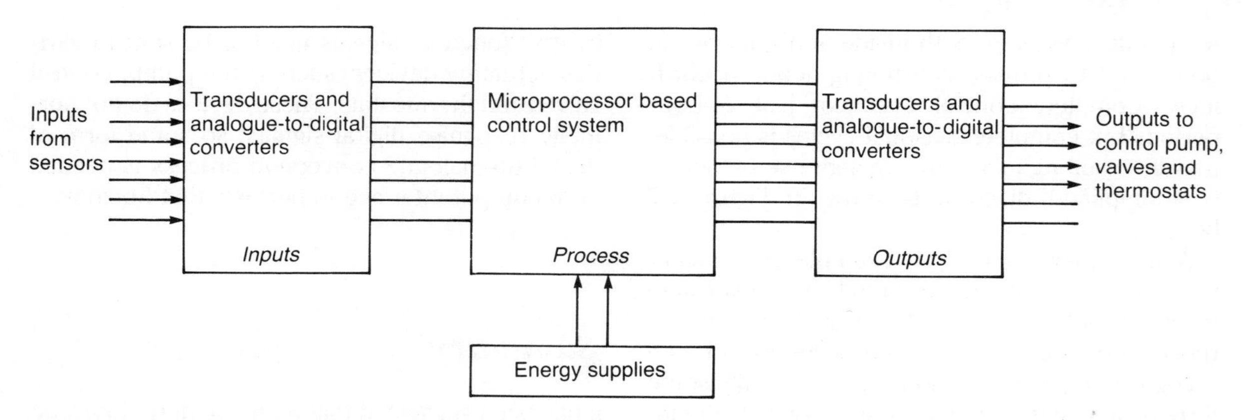

Figure 1.8 Microprocessor based controller system

Transducers To convert the real signals to those the system can recognise and process.
Inputs All of the external signals the system should respond to.
Process A form of changing the input signals in some way to achieve the desired output state.
Outputs The signals resulting from the effects of the processing of the inputs.
Energy Needed to drive the system.

Suggested Assignment Work

1 Explain the functions of the various transducers found in the washing machine.

2 Draw block diagrams of the following items as systems:

a radio set
a sewing machine
a motor car
a lathe
a power station.

3 From your own experience suggest what types of transducers are most commonly found in everyday life.

TWO

Flow Charts

OBJECTIVES

At the end of this unit all students should be able to:

Deduce a flow chart for a typical controller sequence (e.g. washing machine, central heating system) and hence explain the sequence of events that occur within the system.

Once an item or artefact has been drawn as a system in block diagram form it will be of little use to a systems engineer. What is needed now is some structured way of describing the process or processes which are taking place inside the block labelled 'processes'.

Using the examples of electrical appliances from Chapter One it will be shown that each item or appliance, which is now being treated as a system, may have its overall function more clearly understood and described more fully by breaking it down into smaller parts. If each of the smaller parts can be understood and described simply, then we are more likely to understand the complete system.

Time must come into our understanding of the operation of the system. This is where flow charts are useful and the symbols we use are shown in Figure 2.1 later on. Flow charts are ways of showing a series of events, happening one after the other, indicating what the system does when it is working properly. Examples are shown later on in this chapter but as an illustration now consider a cassette player we would like to use. We would check the batteries, switch it on, insert a cassette, set it to play, adjust the volume: a reasonable series of events or steps to take to listen to a tape playing.

These individual steps can be described by using a flow chart. This shows the process broken down (or reduced, or refined as it is sometimes called) into smaller stages.

To do this successfully a standard system of flow chart symbols has to be used. These symbols must be simple enough so that their *shape* conveys a visual image of their meaning and be relatively few in number so that interpretation problems can be eliminated.

The standard symbols that will be adopted throughout this book are shown in Figure 2.1 below.

The symbols that are most commonly used are those of input/output, process and decision. The symbol sizes have to be large enough to allow for

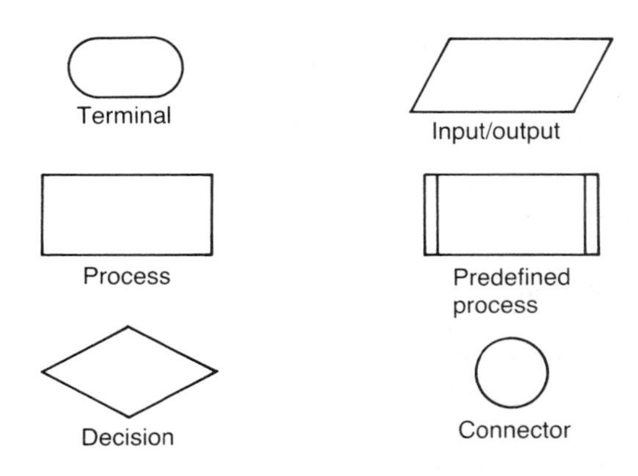

Figure 2.1 Symbols used for flow charts

a short sensible comment which will be meaningful to the person reading it. Before examples of flow charts are developed, it is appropriate at this point to introduce the concepts of *sequence*, *selection* and *iteration* which are some of the terms commonly used in structured programming. This is a way of writing computer programs in such a way that they have the following properties:

ease of reading
ease of maintenance (development)
clear and logical progression
each module has single entry and single exit

It should be recognised from the earliest point that software, like hardware, has a life-cycle of its own. Just as a hardware item like a washing machine, say, or a compact disc player is designed, developed, prototyped, tested and produced so also is software.

It pays from the very start of dealing with microprocessor systems or computer programming to get into good habits.

The three ideas of sequence, selection and iteration will be investigated from the point of the flow chart symbols available.

SEQUENCE

As the name suggests, this implies a series of events following a logical order. Most of these could be called applied 'common sense' but sometimes in an engineering situation the exact sequence could be difficult to define, or to observe.

To give a couple of examples:

The process of getting up in the morning could include the following typical events:

Alarm rings
Switch off alarm
Get out of bed
Go to bathroom
Get dressed
Have breakfast
Listen to radio
Watch TV
Read paper
Leave house

The order of these events listed above is a typical order, but is not the only one.

For example:

Get out of bed
Switch off alarm

could be a viable alternative.

However, it should normally be very difficult to do the following:

Go to the bathroom
Get out of bed

Also it is normal to get dressed *before* leaving the house!

The list includes key points or *milestones* in the process which, at some point or other, will be followed by the program:

i.e. Get out of bed
Get dressed
Leave the house

The *order* in which the other processes are followed is defined either by personal choice, common sense or elementary hygiene!

The flow chart symbol used would be that of *process* as shown in Figure 2.2.

The comments have been listed inside each process block. Some programmers like to write the comments *alongside* the process blocks, as quite often space does not allow this.

An engineering example which is quite simple is to look at what happens when a switch is closed in a simple circuit as shown in Figure 2.3.

When switch S is closed, the lamp lights! As a process this is obvious.

Unfortunately, the process is not as simple as it looks. Although the switch is not heavy, the contacts usually do not close cleanly and show what is called 'contact bounce'.

If the situation could be imagined where a rubber ball is dropped from a height of 1 metre onto a concrete floor, it is to be expected the ball will bounce back up. This process will continue until the ball stops bouncing as shown by Figure 2.4.

A similar thing happens with switch contacts and the graph could be redrawn as shown below in Figure 2.5.

Eventually the switch is closed and the current can flow continuously. How long does this process take? It has nothing to do with 'common sense'. The solution to the switch bounce problem is usually solved by waiting a short period of time before testing the value of the current flowing in the circuit.

In this case the sequence could be as shown in Figure 2.6.

The time involved is usually of the order of milliseconds but it can be evaluated by examining the behaviour on an oscilloscope.

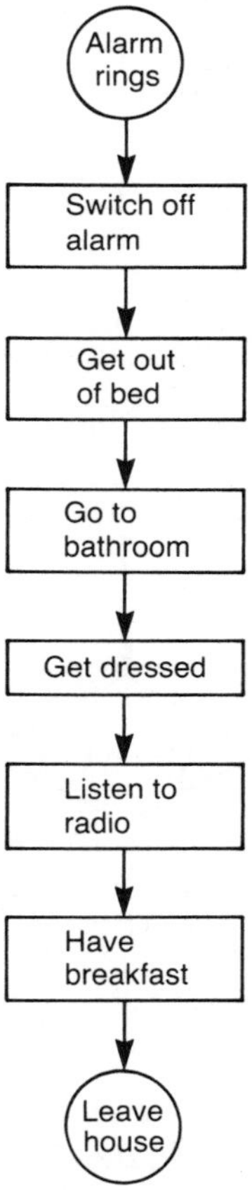

Alarm rings
Switch off alarm
Get out of bed
Go to bathroom
Get dressed
Listen to radio
Have breakfast
Leave house

Figure 2.2 Flow chart for getting up in the morning to show 'sequence'

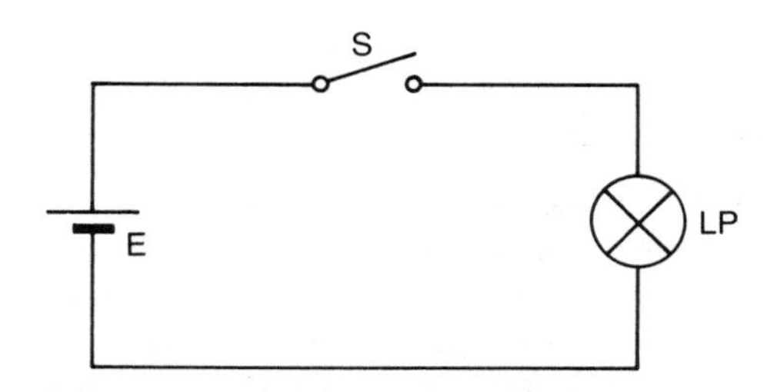

Figure 2.3 Light controlled by a switch

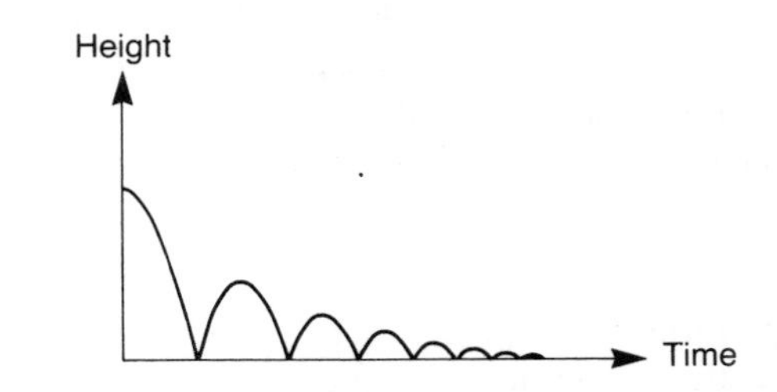

Figure 2.4 Bouncing ball losing height as time passes

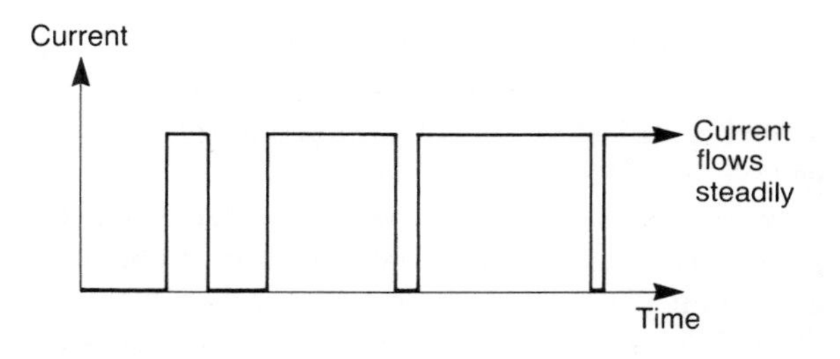

Figure 2.5 Effect of switch contact bouncing

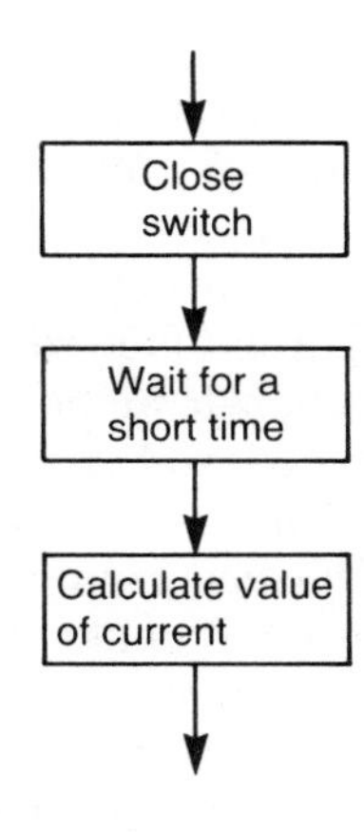

Figure 2.6 Flow chart of process to solve contact bounce problem

SELECTION

Quite often a sequence of steps in a process is carried out until a point is reached where a *decision* has to be taken. Depending upon certain circumstances, the program can then *branch* elsewhere. The simplest situation here is a straightforward choice from *two* possibilities but others can exist which will not be dealt with here.

To return to the process of getting up in the morning, there exists the possibility of choice and hence a decision can be made. Assume that breakfast is being eaten and that a choice can be made as follows:

Listen to the radio or not?
Watch the TV or not?
Read the paper or not?

Here, three choices are possible. How could these be handled in a flow diagram? Figure 2.7 shows one method.

Here the *selection* is made by repeated use of the *decision box*. It is a very useful symbol to have available and its use is not confined to *selection* as will be seen later.

When following the program, a sequence of *selections* are made, and the operator of the program can choose whether to have breakfast in total silence or have radio and television on and competing for attention.

In the flow chart shown in Figure 2.7 it is seen that by following the program from the top of the page downwards a point is reached where a decision box is met. A question is posed by a *statement – 'Listen to radio?'* The response is either 'yes' or 'no' and the decision box is left at one of two possible exits depending on the decision made. The 'yes' exit is followed by a process to initiate the action chosen, in this example, 'switch on the radio'. The 'no' exit bypasses that process. The other choices follow on, one after the other as the flow chart is followed from top to bottom.

This type of flow chart could also be applied to a bouncing switch as shown by Figure 2.8.
In this flow chart, the selection process has to detect, by a method unspecified here, if the contacts are properly closed or still bouncing about. In this case, the decisions are as before, 'yes' or 'no', with the flow chart showing how the operation should work.

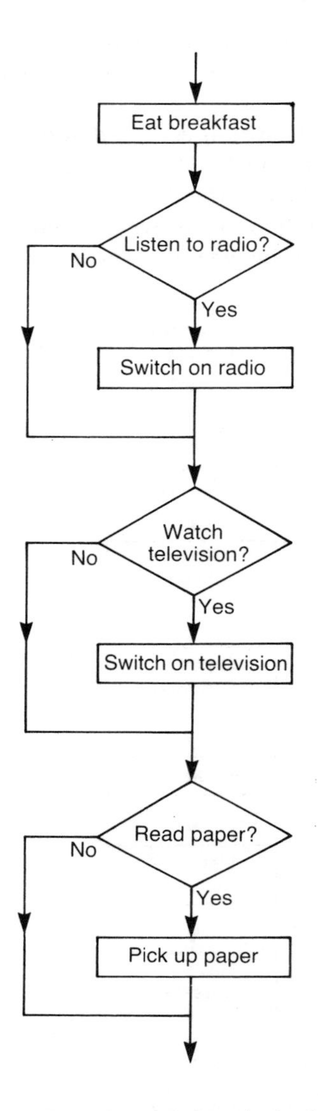

Figure 2.7 Flow chart to show 'selection'

ITERATION

This is the method of *repeating* a process until a *satisfactory outcome* is reached. This usually involves *decision* boxes also! To explain what is meant by a *satisfactory outcome*, consider again the flow chart for getting up in the morning.

Most people need to put on *two socks* and *two*

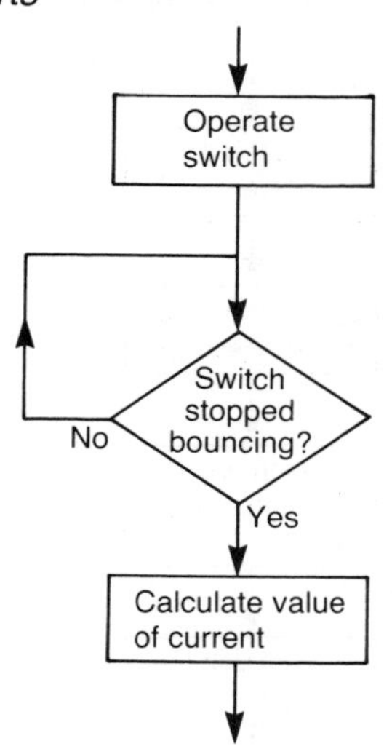

Figure 2.8 Flow chart showing 'selection' applied to a bouncing switch

shoes as part of the dressing process. A computer program would need to be told how to do this and the flow chart could look like Figure 2.9.

Here the decision process involves the idea of a *counter*, often called a *loop counter*.

This idea could also be applied to the bouncing switch contacts. Let it be assumed that it takes 10 milliseconds* for the switch contacts to stop bouncing about. Figure 2.10 shows this.

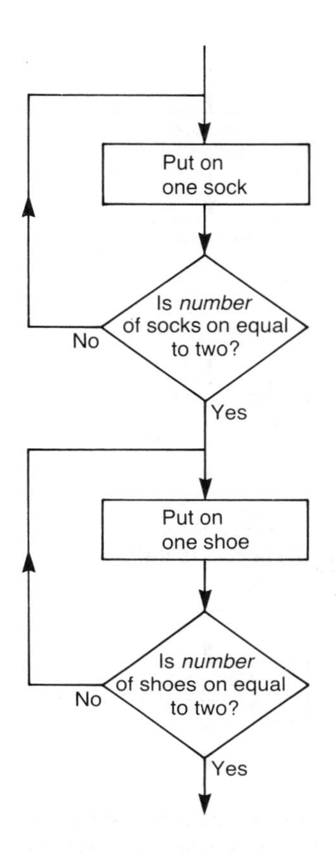

Figure 2.9 Flow chart showing 'iteration'

---

* Engineers and Scientists use a series of terms and units, called the SI system [SI stands for System Internationale] which uses the metre as the unit of length, the kilogram as the unit of mass and the second as the unit of time. The kilogram (kg) is 1000 grams and so 'kilo means 1000 of'. This is abbreviated, as shown in the brackets, to 'kg'.

There are other multiples and dividers used as follows:

T	=	tera	=	10^{12}	or	1 000 000 000 000 times
G	=	giga	=	10^{9}	or	1 000 000 000 times
M	=	mega	=	10^{6}	or	1 000 000 times
k	=	kilo	=	10^{3}	or	1000 times

We can also have multipliers which make things smaller, or dividers. These can be used to describe small quantities such as short intervals in time.

m	=	milli	=	10^{-3}	or	$\frac{1}{10^3}$	or	$\frac{1}{1000}$	or	0.001
μ	=	micro	=	10^{-6}	or	$\frac{1}{10^6}$	or	$\frac{1}{1\,000\,000}$	or	0.000 001
n	=	nano	=	10^{-9}	or	$\frac{1}{10^9}$	or	$\frac{1}{1\,000\,000\,000}$	or	0.000 000 001
p	=	pico	=	10^{-12}	or	$\frac{1}{10^{12}}$	or	$\frac{1}{1\,000\,000\,000\,000}$	or	0.000 000 000 001

So 10 milliseconds (10 ms) = 10 thousandths of 1 second.

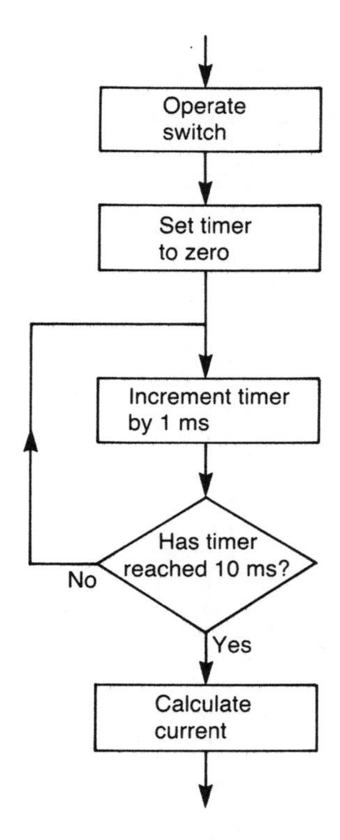

Figure 2.10 Flow chart showing 'sequence', 'selection' and 'iteration' applied to a counter

Here it can be seen that by initialising (which is a term used to set a counter to a required initial or first value) a loop counter to zero followed by a process to increment the counter by one every millisecond can be followed by a decision box which will only be satisfied after a time of 10 ms.

These three concepts of sequence, selection and iteration will be used many times in problems in real life and will be used again in the examples in this book. In the rest of the chapter simple flow charts will be produced for the systems described in Chapter One.

PRODUCING FLOW CHARTS FOR SYSTEMS

Cassette Tape Player

Figure 2.11 shows a flow chart to operate a cassette tape player using a pre-recorded tape.

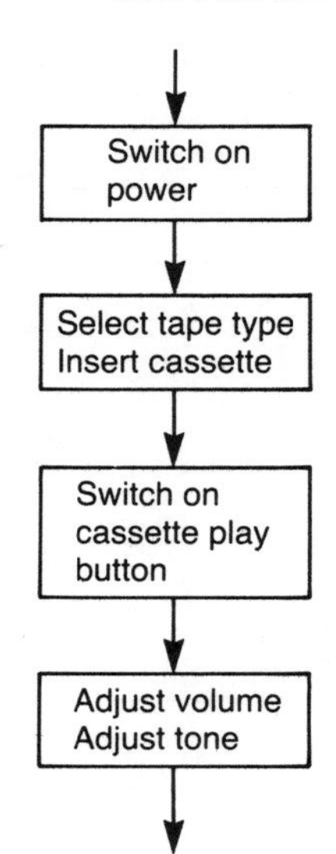

Figure 2.11 Flow chart to operate a cassette player

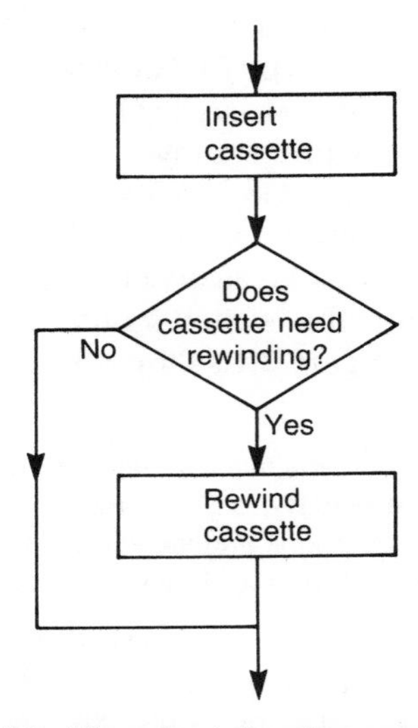

Figure 2.12 Flow chart showing process to rewind a cassette tape

Should the cassette need to be rewound, then Figure 2.12 shows how this could be represented in flow chart form.

Microwave Oven

Figure 2.13 shows a flow chart to operate a microwave oven safely. The oven cannot operate until the door is safely closed.

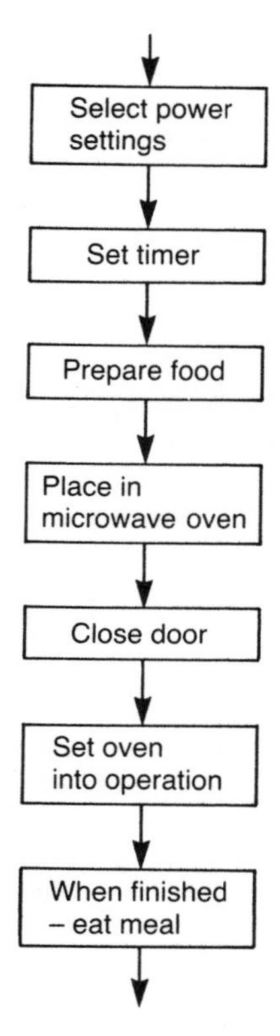

Figure 2.13 Flow chart to operate a microwave oven

Washing Machine

Figure 2.14 shows a flow chart to operate a washing machine, the processes following each other in sequence.

Once again, each of these process blocks can be represented by further refinements:

e.g. the 'fill up with cold water' block in Figure 2.14 could be as shown by Figure 2.15.

Central Heating

Figure 2.16 shows a possible flow chart for a central heating system. This system needs to be able to detect differences between ambient temperature in living rooms and the much higher temperature needed for hot water.

A flow chart for such a system could be drawn as shown in figure 2.17

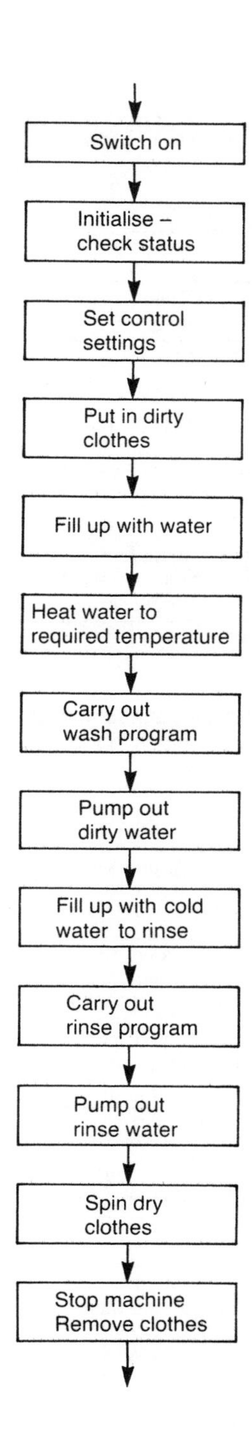

Figure 2.14 Flow chart to operate a washing machine

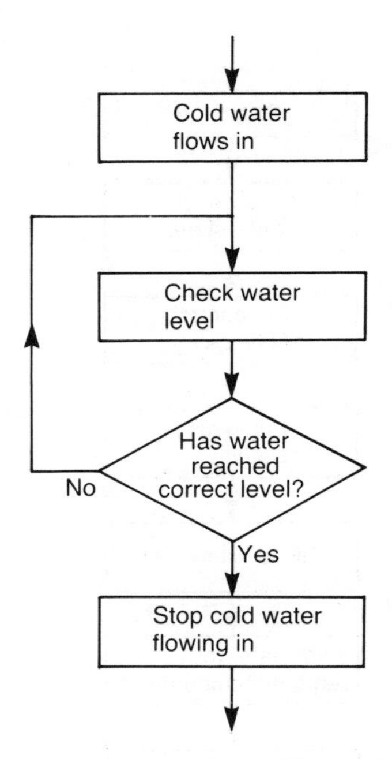

Figure 2.15 Partially refined flow chart to replace the 'fill up with water' block in Figure 2.14

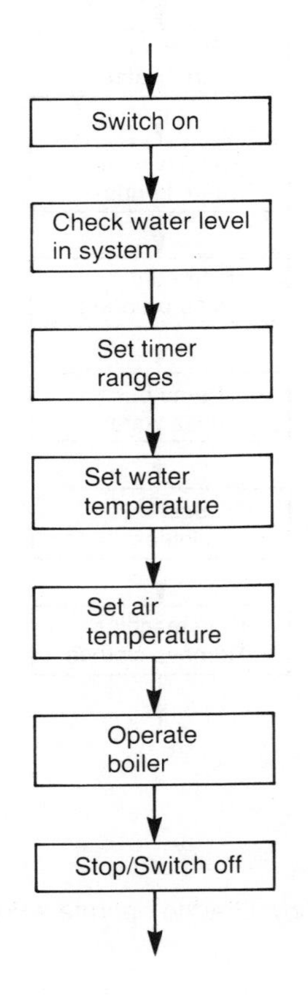

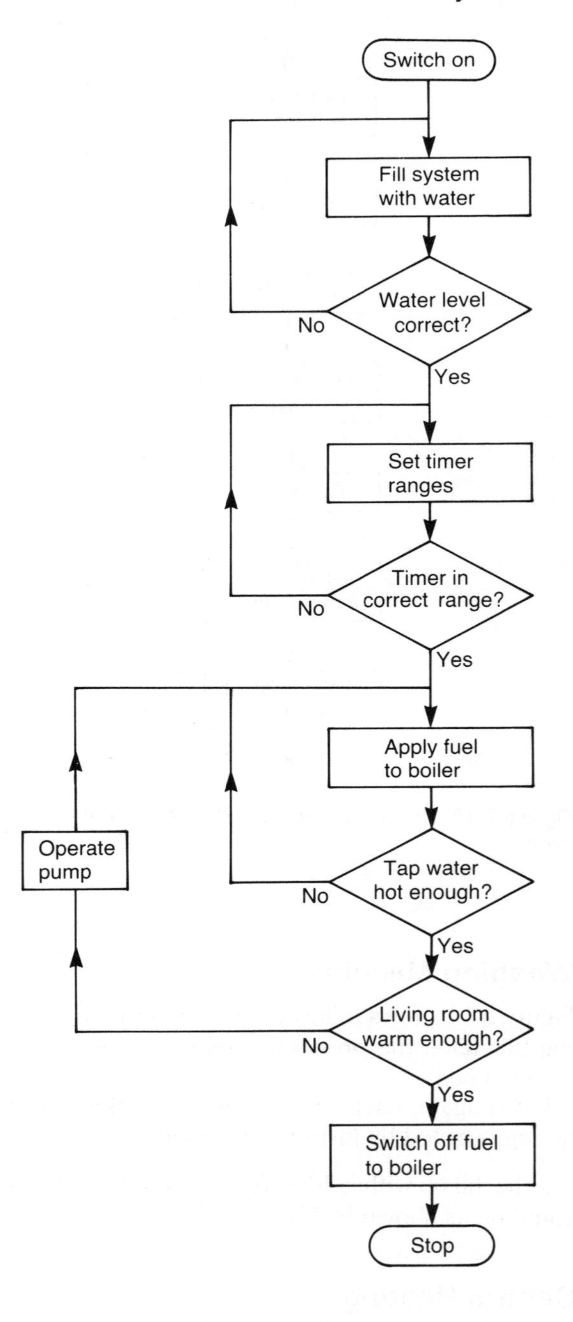

Figure 2.17 Improved flow chart for operator of central heating showing 'sequence', 'selection' and 'iteration'

← **Figure 2.16** Flow chart for operator of central heating

Compact Disc Player

This is a similar system to the cassette player but has many more advanced facilities because of the way a microprocessor controls the laser which reads the disc.

Figure 2.18 shows a flow chart to operate a compact disc player.

The disc itself has a directory track which tells the laser reader how many tracks there are on the surface of the disc and the duration of each track. The person playing the disc can program the player to play each of the tracks in any order of their choice, or in default of a program, to play the disc sequentially. Shuffle or repeat programs are possible.

Also, more importantly, the sound track is laid down on the surface of the disc by using a digital recording technique. In this case the sound is sampled at a high frequency and the instantaneous value is converted into a digital form. On playback the reverse takes place. The digital pattern is passed through a digital-to-analogue (DAC) conversion process and passed through a low-pass filter before being amplified. (A low-pass filter is a frequency selective device which will allow through it signals which have frequencies ranging from very low values up to a certain value called the critical frequency. Above the critical frequency, a low-pass filter attenuates the signals very greatly, or reduces their amplitudes so that we think those frequencies are missing. Amplitude is described more fully in Chapter Three but is a way of describing the size or magnitude of an alternating waveform.)

The flow chart would, in principle, look very similar to that of the cassette player, but be modified to take into account the microprocessor control system.

Figure 2.19 shows how the 'Select and play' program could be shown in flow chart form.

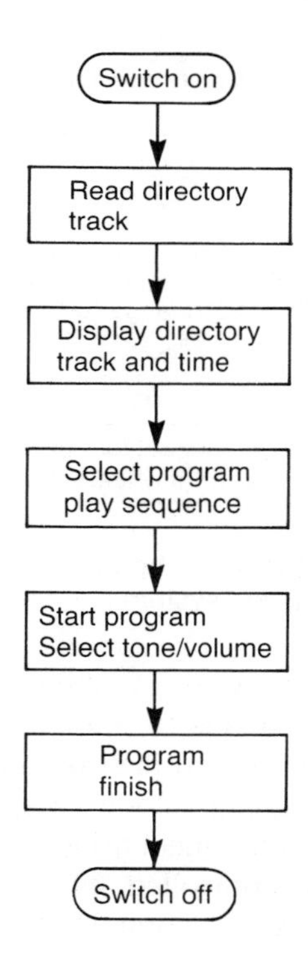

Figure 2.18 Flow chart to operate a compact disc player

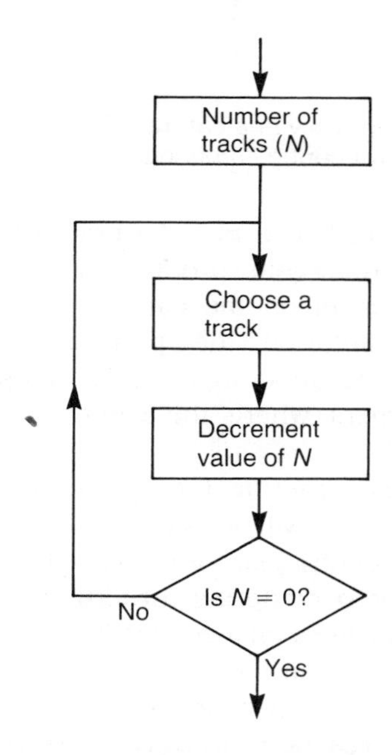

Figure 2.19 Improved flow chart to replace 'select program play' sequence in Figure 2.18

Active Suspension System

Modern racing cars and some of the more sophisticated road cars use an active suspension system. This has been developed during Grand Prix

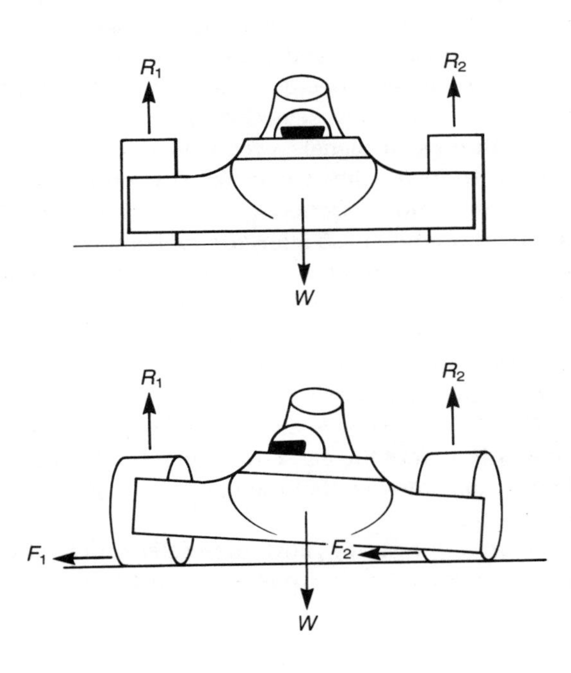

Figure 2.20 Front view of a car travelling (a) in a straight line and (b) during cornering

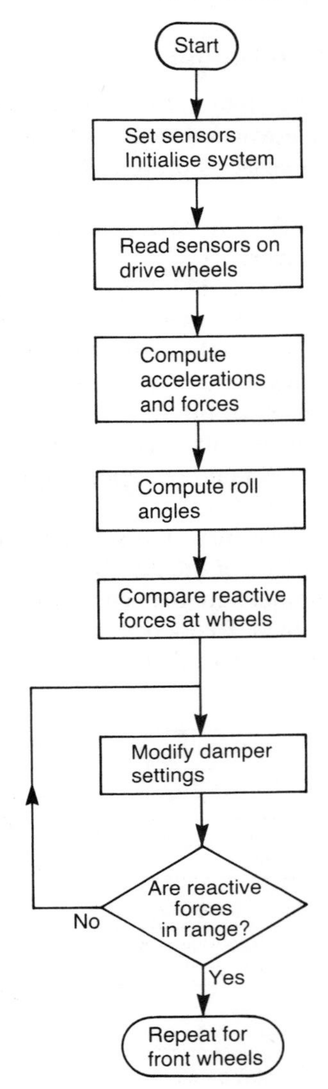

Figure 2.21 Flow chart showing operation of an active car suspension system

racing on Formula One cars and is gradually filtering its way into road-going production cars. In this system the sensors on the car suspension are continually monitoring the performance and attitude of the car on straight roads, bumpy surfaces and corners where the camber changes drastically.

When a car is cornering, the reaction forces R_1 and R_2 vary. The car will not slide or skid if F_1 and F_2 balance the centrifugal force. With the change in attitude of the body during cornering R_1 and R_2 are unequal causing a change of driving torque to the wheels which can cause loss of traction, or skidding. An active suspension unit senses the changes in the reaction forces (R) and frictional forces (F) and modifies the stiffness of the struts or shock absorbers to reduce the vehicle roll angle. When cornering, the vehicle theoretically should adopt an attitude more like that shown in Figure 2.20(a).

The sensor unit is continually checking the balance of forces at each of the wheels in turn but is likely to be checking the effects on the driving wheels at least twice as often as the steering wheels. (Whilst this is true for Grand Prix racing cars, the system could also be adopted for conventional road-going saloon cars, which may, or may not, have rear wheel drive.)

It is usually controlled by a microprocessor system. Sensors are connected in both vertical and horizontal planes, giving three-dimensional sensing. Accelerometers can be used to detect *change* in velocity. Computer programs can then compute the values of the various forces at work and the velocities involved and send signals to the suspension units which change the damper rate settings. Figure 2.21 shows how a flow chart could represent the processes in an active suspension system.

Summary

A range of quite widely differing systems have been represented in block diagram form in Chapter One.

These block diagrams should show the outline of the system but not necessarily the steps the process actually follows.

By using standard flow chart symbols and the concepts of *sequence*, *selection* and *iteration* it is seen that each step in the process block diagram can be represented by a flow chart. Each flow chart shows how the process actually follows a logical progression of events to make the system effective. Some systems are obviously more complex than others, but the flexibility of the flow chart system allows the system designer or system engineer to portray his system and its parameters to others.

The flow chart symbols most commonly used are shown in Figure 2.1 on page 7.

Suggested Assignment Work

1 Suggest the features or functions of the various transducers in one of the systems described in this chapter.

2 Take some of the process blocks in the central heating system or the washing machine and apply the process of refinement further. This means deciding for yourself what type of transducer and control device you would use to carry out the function implied by that decision box and process block.

3 From library material investigate how the compact disc player carries out the processes described.

THREE

Analogue and digital signals and systems

OBJECTIVES

At the end of this unit all students should be able to:

Demonstrate the difference between analogue and digital signals and systems with reference to:
the parameters of an analogue signal
the parameters of a digital signal
the factors involved in signal conversion
examples of analogue and on/off transducers.

The real world in which we live, work and enjoy ourselves is an analogue world. By that it is meant that events happen and change in a relatively continuous manner. When change occurs it is quite often continuous in nature, even though the rate of change involved can be very rapid, even frantic.

Attitudes of people living in industrialised societies have changed drastically since just before 1950 when the transistor was discovered. This has led to the development of the modern microelectronic systems that are around us today and that are increasingly taken for granted.

The transistor was initially used in amplifiers and other signal shaping circuits in functions that would now be termed 'linear' or 'analogue'. It should not be forgotten that at the same time engineers were developing transistors to act as switches, to control signals by turning them on and off. This led to what is termed 'discrete' or 'digital' functions. The two technologies have developed side by side over the years in a seemingly frantic race. First one technology sees drastic developments and then the other. Each has its adherents and supporters but the complete engineer or technician engineer should be versatile in either technology and, more importantly, use the correct technology in the right way.

As microelectronics developed in both analogue and digital areas, so the pace of change accelerated. Modern transport systems are increasingly complex and rapid. Consider as examples the driverless trains in many cities of the world, high speed trains in use in Japan and France and supersonic air travel in Concorde.

All this, of course, is fed by developments in the military sphere where research and development are seeking ever more rapid aircraft, 'smart' weapon systems and the development of countermeasures against enemy 'high-tech' weaponry.

The important single feature about all of this is the *integration* of analogue and digital electronics into complex systems relying heavily on microprocessor or computer control because their response times to outside events are often far too rapid for humans to handle.

An understanding of the nature of analogue and digital signals, the differences between them and the conversion between them is of vital importance in any application of microelectronics. Without this understanding, new systems will fail to be designed, developed, used and maintained effectively. All of these roles are important if overall success is to be obtained.

Let us look at the different features of analogue and digital signals, see where the differences lie and then develop methods for converting between one system and the other.

The next part of this chapter will introduce some of the terms used by electronic engineers to describe and illustrate both analogue and digital signals. You will meet terms like alternating signals, sine waves, sinusoidal, cycle, period, periodic time, amplitude, frequency, phase, phase angle, phase difference, voltage, hertz, degrees, root mean square values, logic 0, logic 1, etc.

You should have met these terms before this, but it may be useful for you if they are briefly described again as they appear in the text over the next few pages.

If you refer to Figure 3.1 you will see the graph of a *sine wave*. We call its shape *sinusoidal*. The shape comes from trigonometry but it is important to electronic engineers for a number of reasons, including:

(a) Many analogue signals (remember from Chapter Two, these are the continuous ones) are alternating signals. This means that they can have values which change with time, and can be either positive or negative at certain instants in time. This is shown on the graph as being either above or below a horizontal line called the *horizontal axis*.

(b) Many analogue signals repeat themselves at regular intervals, and we call this behaviour cyclic, i.e. going in cycles. A natural example of a cycle is the tidal cycle in which the tide ebbs and flows twice a day in a predictable fashion. This leads to the idea of the time taken for one complete cycle, called *periodic time*, or often just *period*. Natural periodic times include 28 days for the moon to orbit the earth (lunar cycle) and 365 days for the earth to orbit the sun.

(c) If these analogue signals are going through their cyclic variations in a very rapid manner then the period could be very short, often less than one second in duration. This leads to the idea of how many cycles could take place in one second and this is called frequency.

FREQUENCY is the number of *cycles per second*
PERIOD is the number of *seconds per cycle*

Frequency is measured in hertz (Hz)
Period is measured in seconds (s)

The unit of frequency is named after the famous pioneering physicist, Heinrich Rudolph Hertz (1857–94).

(d) These signals can alternate in a regular, cyclic, repetitive and predictable manner. If they do it is likely they are behaving in a sinusoidal way. This means that if we plot the actual behaviour of these signals on graph paper then they would look just like the sine wave shown in Figure 3.1. This has been found to be true in so many examples that the sine wave is treated as a very important tool of our trade and is used to describe analogue signals mathematically.

ANALOGUE SIGNALS

We can consider the case of the portable cassette player, the 'Walkman' that has become used by young and older people alike because of its compact size and quality of sound.

Audio specifications have changed over the years. Equipment can now provide quality audio over a very wide frequency range, from a few hertz to well over 20 000 Hz. It can also operate over a wide dynamic range which means that the equipment can handle very low level signals in quiet passages to the very high level signals in loud music.

This tells us something about the way in which electronic noise in equipment has been reduced over the years. You only have to listen to modern digitally mastered CD systems to appreciate this.

One factor which has remained constant throughout this development is the capability of the human ear to appreciate the technical finery available! As people get older their audio frequency range does deteriorate but that does not prevent their seeking equipment that delivers a signal with a good dynamic performance over a wide frequency range.

So, what are the properties or parameters of analogue signals? Let us consider a cassette system with a frequency range of 20 Hz to 16 kHz.

This means that the *bandwidth* of the audio

amplifier in the player lies between 20 Hz and 16 kHz (the k stands for 'kilo' meaning 'thousands of . . .').

All of the signals that we are talking about are examples of *sine waves*. *Sine* is a function that is used in trigonometry, but here we use it to help us understand an electrical signal. A sine wave is shown in Figure 3.1 below.

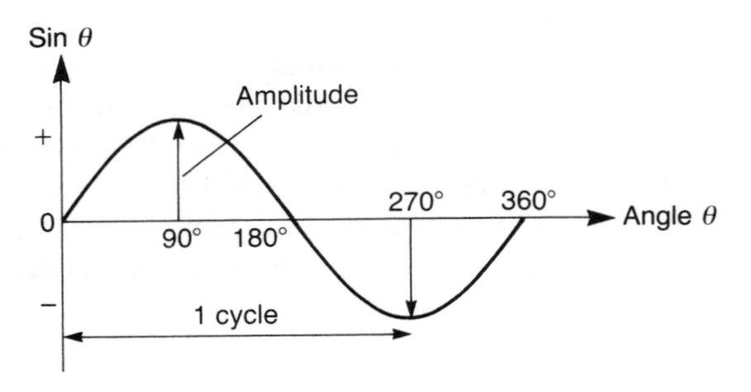

Figure 3.1 Graph of sin θ

If you use an electronic calculator to tell you the values of sin θ (θ is the Greek letter 'theta', which is often used as a symbol for an angle) for angles between 0° and 360° you will see they range between the values in Table 3.1 below.

Table 3.1

Angle θ	Sin θ
0	0.000
45	0.707
90	1.000
135	0.707
180	0.000
225	–0.707
270	–1.000
315	–0.707
360	0.000

The amplitude of sin θ is 1.000. This is the value of the maximum positive point.

You will notice that the sine function has values which can be positive or negative, depending on the value of the angle θ. Between 0° and 180° the sine function is positive. This means the graph when plotted lies above the horizontal axis. Between 180° and 360° the sine function is negative. When plotted this part will lie below the axis.

Notice also that when $\theta = 90°$, the sine function has its maximum positive value equal to 1. This is given the special name of *amplitude*. So the amplitude of sin θ is 1.

The graph of sin θ is an important wave shape because it applies to the many signals in electrical and electronic engineering, including the 240 V, 50 Hz a.c. mains supply.

Let us stop and examine that statement about the mains supply; 240 V, 50 Hz – what does that mean and what has it got to do with a sine wave?

If you 'see' the mains supply displayed on a cathode-ray oscilloscope (CRO), a very useful piece of test equipment in both analogue and digital electronics, then you will see something like the shape shown in Figure 3.2.

The sine wave keeps repeating itself every 20 ms, so a question which needs to be answered is: 'what is the relationship between angle in Figure 3.1 and time in Figure 3.2?'.

Frequency

The tie-up between angle and time is *frequency*. Frequency (f) is measured in *hertz* (Hz), another name for cycles per second, and one cycle represents 360°. So 50 Hz means 50 cycles per second. Further, 50 Hz means 1/50 second for one cycle or 0.02 seconds.

This is known as the *periodic time* T and $T = 1/f$

The *larger* the value of the frequency (f) the *smaller* the value of the periodic time T.

Amplitude

If you look at Figure 3.2 again you will notice that the vertical axis is in *volts* and the peak or maximum value of the sine wave is 340 V. This is much larger than 1, the amplitude of sin θ mentioned earlier on this page. Because the sine waves of actual signals may have maximum values larger or smaller than 1, we use the term *amplitude* as a multiplier for all points on the sine wave.

The amplitude of the UK a.c. mains supply is actually 340 V. If you plot the waveform, you will

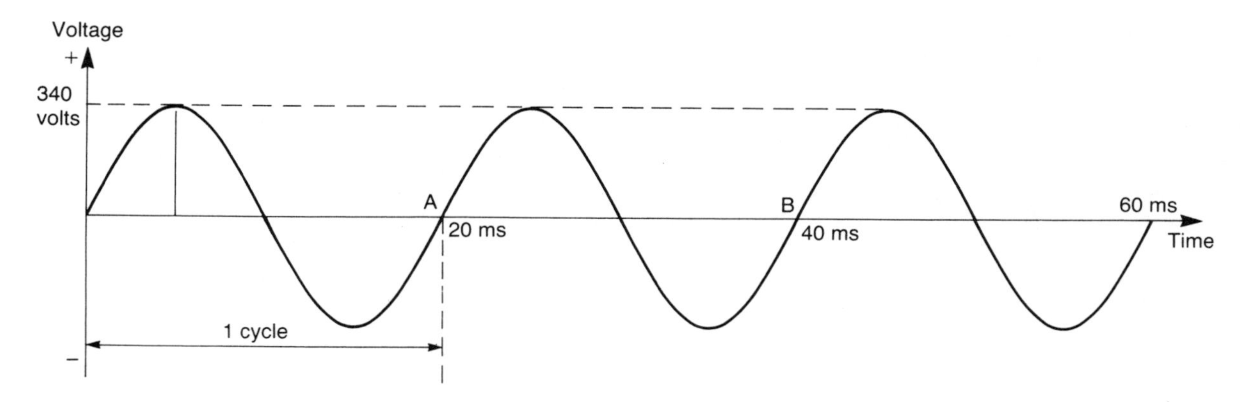

Figure 3.2

find it is sinusoidal in shape but scaled up 340 times. The point values are shown in Table 3.2 below:

Table 3.2 How an amplitude of 340 multiplies the point values on a sine wave

Angle θ	Sin θ	340 × Sin θ
0	0.000	0.000
45	0.707	240.380
90	1.000	340.000
135	0.707	240.380
180	0.000	0.000
225	–0.707	–240.380
270	–1.000	–340.000
315	–0.707	–240.380
360	0.000	0.000

Electrical engineers usually describe alternating waveforms in terms of 'root-mean-square' (r.m.s.) values which has the effect of comparing them with d.c. or non-alternating quantities. These r.m.s. values are used to help engineers compare the power available (particularly heating effect) from sinusoidal signals like the a.c. domestic mains supply with that available from d.c. supplies such as batteries. For a sine wave the r.m.s. value is calculated by measuring the *amplitude* or maximum value and dividing it by $\sqrt{2}$ or 1.414.

So $\frac{340\text{ V}}{\sqrt{2}} = 240$ V, the figure quoted for the domestic a.c. mains supply, which is the reason we talk about a 240 V mains supply.

Phase

Another term that is used to describe analogue signals is that of 'phase'. You may have heard the term in a different context, e.g. 'the next phase of the project is to . . .' The term implies something to do with timing.

When we examine two sine wave signals of the same frequency on an oscilloscope (CRO) we may see something like that shown in Figure 3.3 on the next page.

Waveform A starts at point O whereas waveform B starts at point O′ which is later in time than O. As A and B have the same frequency, this *phase difference*, as it is called, is constant and it can be treated as an angle ϕ or as time, as will now be explained.

From the diagram of two sine waves having the same frequency value in Figure 3.3, you will see that the *phase difference* between them can be expressed in units of time, because time is the scale on the horizontal axis. Phase difference is a measure of the difference in the starting times of two sine waves with the same frequency. We can also treat it as an angle if we know the connection between angle, frequency and time.

Figure 3.1 shows that 1 cycle of a sine wave is spread over 360°, and Figure 3.2 shows that 1 cycle of a sine wave representing the a.c. domestic mains supply lasts 20 ms. So 1 complete cycle is equivalent to 360° and takes 20 ms. (It takes 20 ms because the frequency is 50 Hz or 50 cycles per second. So 1 cycle takes 1/50 second,

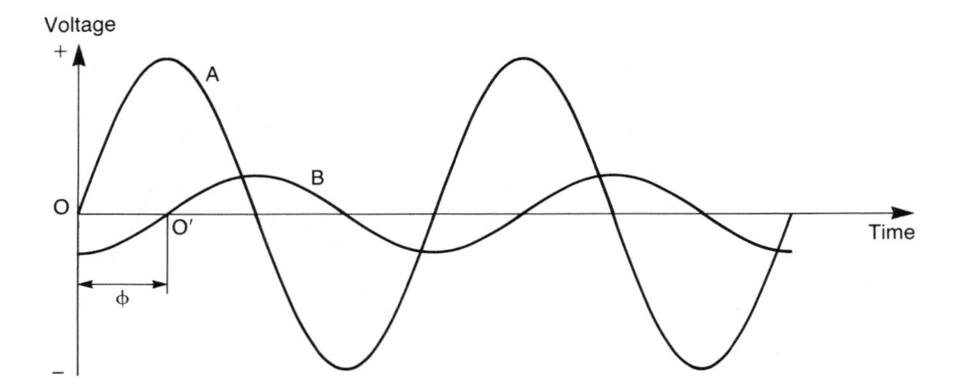

Figure 3.3 'Phase difference' between sine waves of the same frequency but starting at different times

which is 20 ms. Note that the higher the frequency value the shorter the cycle time taken.) This can be scaled up or down as follows:

1 cycle	≡	360°	or 20 ms
$\frac{1}{2}$ cycle	≡	180°	or 10 ms
$\frac{1}{4}$ cycle	≡	90°	or 5 ms
$\frac{3}{4}$ cycle	≡	270°	or 15 ms

The machines in the power station generating our electricity can be thought of as turning through one complete cycle, or 360° in every 20 ms. Applying this to phase difference as shown in Fig. 3.3 means that the horizontal axis could equally well be described in angular terms as degrees.

The use of an angle in this way would lead to phase difference being described as an *angular* difference between the two sine waves A and B.

In a complex audio signal, there are very many components, each at differing amplitudes and frequencies and so we have to be careful how we use the term phase. However, in general terms, we can describe a *sinusoidal* signal by the expression:

$$A \sin(2\pi ft + \phi)$$

where A = amplitude
$\pi \approx 3.142$
f = frequency
t = time in seconds
ϕ = phase angle difference in degrees

DIGITAL SIGNALS

We will now examine some digital signals and describe them in terms of parameters which are similar to those such as amplitude, frequency and phase used to describe analogue signals.

A digital signal is called a *discrete* signal. This has nothing to do with discretion as we would use it in the expression 'using your discretion . . .'; in this case it means separate, distinct.

An analogue signal is continuous and can be varying *continuously*. A digital signal, like a train of pulses varies *discretely* and in *discrete* steps.

Another word often used in conjunction with digital is *binary*, meaning 'two states' which we can describe in a number of ways. These two states have been variously described in text- and databooks by such terms as:

True	and False
High	and Low
Go	and Nogo
Present	and Absent

Another way of description is to use the terms:

Logic 1 and Logic 0

or simply

1 and 0

The word 'Logic' is used to imply a digital state rather than an analogue one. Binary digits, or *bits* for short, are components of digital signals.

In Figure 3.4 opposite a digital pulse train is shown.

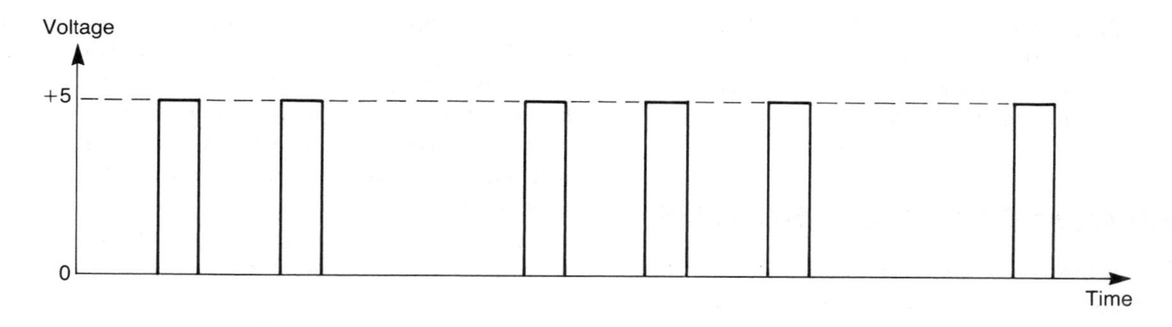

Figure 3.4 A train of digital pulses. Pulses are either present or absent

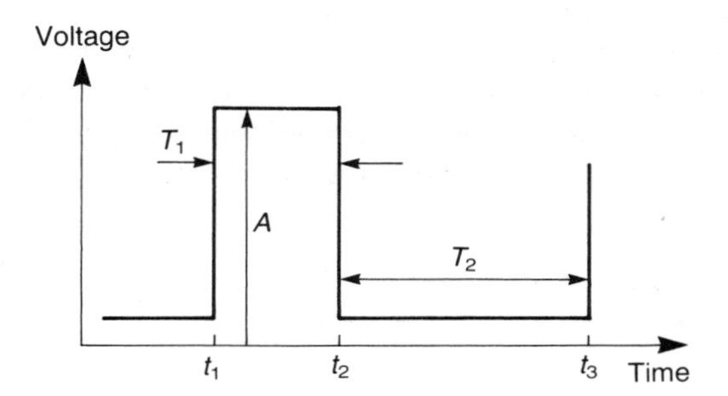

Figure 3.5 Parameters of pulses

Figure 3.6 Pulses described as 'logic 1' when present and 'logic 0' when absent

Let us look at one of these pulses (Figure 3.5).

Amplitude This is the height of the pulse, often in volts.

Pulse period This is the time taken between the start of one pulse and the next. It is very similar to *period* in analogue signals.

$$\text{Pulse period} = t_3 - t_1 = T_1 + T_2$$

Pulse repetition frequency (p.r.f.) This is the reciprocal of pulse period, and is very similar to *frequency* in an analogue signal.

$$\text{p.r.f.} = \frac{1}{T_1 + T_2}$$

Pulse width or duration Even if all the pulses, or bits, are present, there is always a space between them. The pulse width is the time that a pulse can exist for.

$$\text{Pulse width} = t_2 - t_1 = T_1$$

Mark/space ratio This is a measure of the time a pulse can exist for, compared, as a ratio, with the time of the rest of the pulse period.

$$\text{Mark/space ratio} = T_1 / T_2$$

There are other terms that are applied to pulses, such as rise-time, sag, synchronous, asynchronous, but these will not be covered here.

Figure 3.4 shows a succession of pulses and it is obvious that some gaps occur where pulses are missing from the train. Let us look at Figure 3.6 which shows the same pulse train but expressed in a different way.

The binary digits are only allowed to have one out of two possible values and it is sufficient to be able to say whether the pulse is present or not.

If the pulse is present we represent it by a *logic 1*.

If the pulse is absent we represent that by a *logic 0*.

Figure 3.6 shows the pulses with the values 0 and 1 placed in appropriate positions and we end up with a pulse train that could be described by the eight bit wide word 1 1 0 1 1 1 0 1. This type of representation of eight pulses is useful in microprocessor technology and a collection of eight bits is usually called a *byte*.

SERIAL AND PARALLEL SYSTEMS

In all of the above descriptions the idea has been conveyed, quite deliberately, that both analogue and digital signals are either continuous or discrete time-varying signals and also that we are looking at signals at one particular point and observing how they change as time changes.

In practice, many systems are not as simple as this and it does not require much thought to think of examples of either serial or parallel systems.

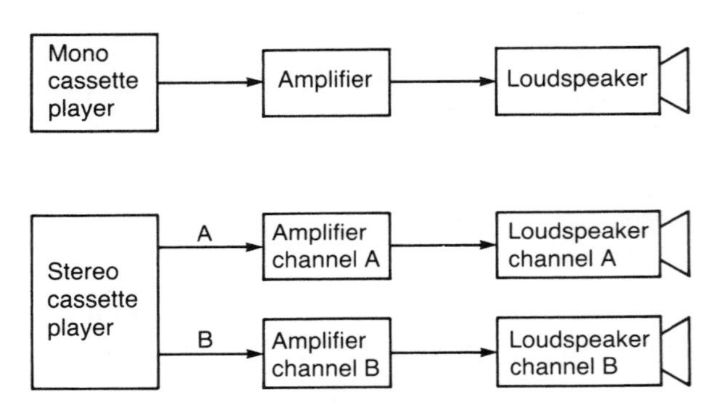

Figure 3.7 (a) A serial analogue system and (b) a serial/parallel analogue system

Figure 3.7(a) shows a block diagram of a mono cassette player. This is an example of a serial system because the process is a series of steps following loading the cassette; the audio signal recorded on the tape is processed, amplified and fed to the single loudspeaker as a single, complete signal. Figure 3.7(b) shows a stereo cassette player. This is an example of a system using both serial and parallel processes. What happens is that two serial systems, such as the mono system above, are used to process, amplify and feed to a speaker two slightly different audio signals *at the same time* or *in parallel*.

In a stereo cassette player the two different analogue signals are processed *at the same time* by the amplifiers and the illusion of the stereo effect is created in the brain of the listener when the differing audio signals from the two loudspeakers are sensed by the ears. Anybody who has experienced the stereo effect from a stereo cassette player, a CD player or an FM stereo tuner will have experienced some form of parallel processing.

Further developments in quadrophonic or 'all around' sound led to further channels being added to this type of system and the audio effects created can be quite startling.

The point, however, is that *in the same time frame* a number of analogue systems are processing slightly different information and the system can be called a parallel system.

Other ideas will occur to you when you consider other aspects of audio, video communications and computer technologies. Figure 3.8 shows serial and parallel digital systems.

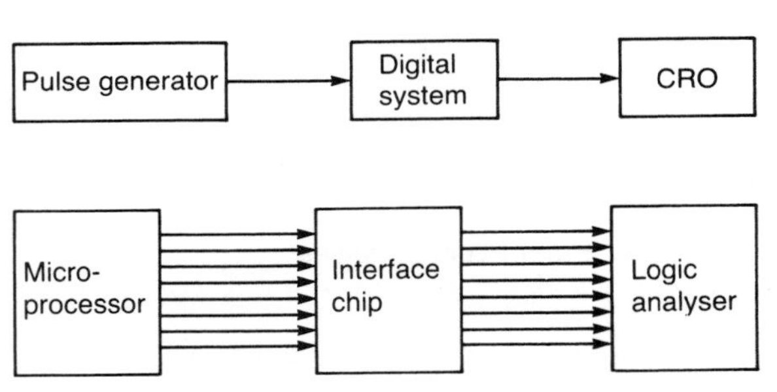

Figure 3.8 (a) A serial digital system and (b) a parallel serial system

In Figure 3.8(a) a pulse generator is being used to stimulate a digital system which will process the *serial* pulse train and the serial pattern can be displayed on a cathode-ray oscilloscope.

Figure 3.8(b) shows a different system. A microprocessor is shown in block diagram format with eight *parallel* data lines coming out of it and going to an interface chip. The main function of this chip is to adjust the pulses for use in the outside world (outside from the microprocessor point of view). When we wish to observe, measure and store such pulse trains for our own use, we use a device called a logic analyser, which can store parallel pulse trains of up to a certain length, or number of pulses. A typical situation would be an eight channel, 16-kilobit analyser, which means that *each channel* can process 16/8 = 2 kilobits of data.

Here we can see that, in *the same time frame*, the microprocessor is sending eight bits of data in parallel to its interface chip, which then passes it out of the system. It does this by using the data bus of the microprocessor system. This is made up from eight wires, all in parallel with each other, connected from eight pins on the microprocessor to an equivalent set of eight pins on the interface chip. These eight wires carry the eight bits of data in parallel mentioned earlier in the paragraph. It is dealt with more fully on page 41.

At this point it would be a good idea to look again at one of our measures or parameters of digital systems. The logic analyser previously mentioned could store eight data lines, each of length two kilobits and it is the measure of kilobits that should now be described.

Up to now the idea of 'kilo' has referred to 1000 or 10^3 times of multiplication, i.e. 1 kilogram = 1000 grams. But in digital terms this is not accurate and we use 'kilo' to represent not 1000, but 1024 times! Why should such a strange number be relevant and why should we want to refer to it as 'kilo' anything?

The reason we use a *kilobit* to represent 1024 bits lies in the value of 2^{10}, which is equal to 1024, and is the closest we can come, using powers of 2, to 1000.

Because semiconductor memories can be built to contain millions of bits it is useful to have a larger unit. So we call, for convenience, 1024 bits 1 *kilobit* (1 kb).

FACTORS INVOLVED IN ANALOGUE-TO-DIGITAL CONVERSION (ADC)

It is apparent that analogue and digital signals and waveforms are quite different from each other. How then can we use a digital processor such as a microprocessor or microcomputer to handle analogue information? The answer lies in converting the analogue information into a digital form which is then acceptable to the digital processing device, e.g. the microprocessor. Once the digital processing has taken place then the results of the process have to be converted back again into an analogue form if it is to be used in the real world. This involves us in the use of *digital-to-analogue converter* (DAC) and analogue-to-digital converter (ADC) chips.

There are various techniques involved in ADC–DAC operations and we will spend a short time looking at the principles and problems of some of the methods used.

The questions that need to be asked in any conversion process are these:

How fast does the process need to be?
How accurate does the conversion need to be?
How does the conversion handle positive and negative signals?
How does the presence of electronic noise affect the process?

It does not matter where you are in the electronic engineering world, or how advanced your own education, training and experience in working with these devices are; the questions still have to be asked and answered!

What will be attempted here is a short description of each question and to put some of the answers into perspective.

How fast?

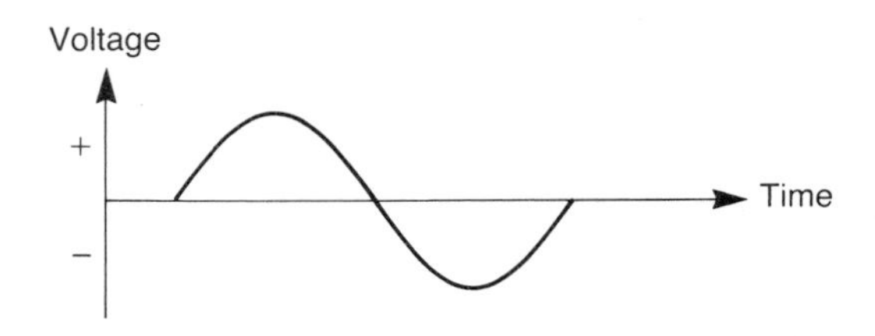

Figure 3.9 A sine wave which is about to be 'sampled'

Or more properly 'How often should the analogue signal be sampled?'

With a sine wave signal as shown in Figure 3.9 we need to have an idea of how rapidly we should *sample* the analogue signal. Should we do it *once* in a cycle, *twice* or many more times than that?

Some years ago Shannon developed an important theory of use to digital electronic engineers. He proved that if the analogue signal could be sampled at a frequency rate *at least twice* as high as the highest frequency component in an alternating signal, then *all* of the information could be extracted. In practical terms, this means finding out the value of the highest frequency component in our signal, doubling its value and using that as the sampling rate.

On page 19, reference was made to quality audio systems capable of reproducing sounds faithfully from disc and tape, which can have frequency components within a range of a few hertz to more than 20 000 hertz (20 kHz). In this case the sampling frequency would have to be 40 kHz, or greater, if all the signal information were to be extracted.

Looking back to Figure 3.9, then, it becomes obvious we need to sample the signal at least twice.

In practice, we often sample at faster rates than that for other reasons.

As a further example, in the modern telephone system which uses pulse code modulation (PCM), the usual audio bandwidth for an analogue channel is between 300 Hz and 3400 Hz. The sampling rate used is 8000 Hz.

The effect of sampling low frequency signals in this way is shown in Figure 3.10 where the sampling rate is much higher than twice the frequency.

The effect of sampling a higher frequency signal in this way is shown in Figure 3.11.

After the signal has been sampled, the resulting value is stored in a special 'sample and hold' circuit, usually made as a chip. The 'sampled and held' waveform as shown is an approximation to the original signal. This does not matter because subsequent processing can remove this peculiar pattern. The important thing to remember is that as long as the sampling is fast enough the final result will not be distorted.

As long as the sampling rate is as specified by Shannon, then all relevant information, such as amplitude, frequency and phase may be successfully recovered and processed faithfully.

How accurate?

Or more properly 'What resolution is possible in a digital system?' A digital process has two states, on and off, 1 and 0, hi and lo, etc. There are many ways to describe these, as shown on page 22.

If you met digital, or binary counting at school

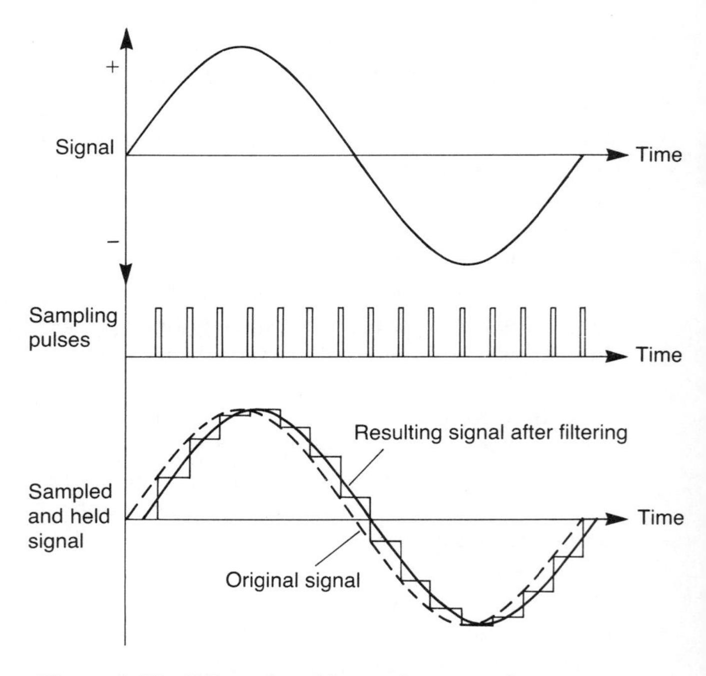

Figure 3.10 Effect of rapid sampling on a sine wave. The resulting waveform is a close approximation to the original signal but slightly delayed

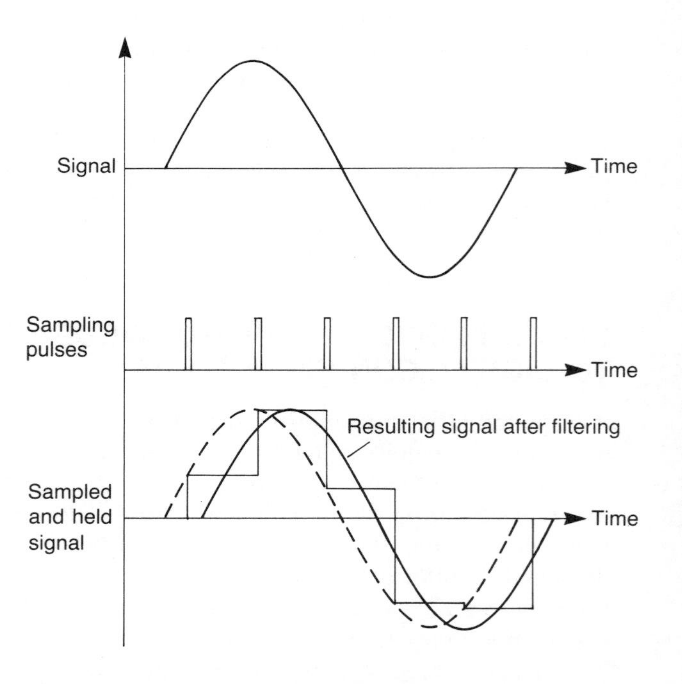

Figure 3.11 Effect of slower sampling. The delay is greater than in Figure 3.10

or elsewhere, you will recognise the table shown in Figure 3.12 comparing denary numbers with their binary equivalents.

In an 8-bit microprocessor, such as the Motorola 6802 or the Intel 8085, an 8-bit *data bus* is used. This term was first described on page 25. This means that *all* processing is carried out eight bits at a time in any one operation. The table above shows a variable number of bits, so we need to know how many denary numbers we can represent with just eight bits.

Denary number	Binary number
0	0
1	1
2	1 0
3	1 1
4	1 0 0
5	1 0 1
6	1 1 0
7	1 1 1
8	1 0 0 0
9	1 0 0 1
10	1 0 1 0

Figure 3.12

The answer is quite simple: if there are N bits, then we can obtain 2^N different digital states as shown in Figure 3.13,

$$\text{i.e. } 2^8 = 256 \text{ states}$$

which means that we can convert 256 denary numbers, which includes zero, into binary form with eight bits. So 0 to 255 can be converted.

Note from Figure 3.13 that the bit position with value 2^0 is labelled 'LSB'. This stands for 'Least Significant Bit' because it has the smallest individual bit value owing to its position ($2^0=1$). Similarly the bit position with value 2^7 is labelled 'MSB', which stands for 'Most Significant Bit' because it has the largest individual bit value owing to its position ($2^7=128$).

In practical terms any analogue signal which has a certain value in a given range, e.g. 0 to 5 volts can be represented by an 8-bit digital number with an *accuracy* or *resolution* of 5 ÷ 256, per step = 0.0195 V ≃ 20 mV. In practice, although only discrete levels are available in digital systems, we can reduce the gap between states by increasing the number of bits available.

Denary number	Binary number							
	2^7 (MSB)	2^6	2^5	2^4	2^3	2^2	2^1	2^0 (LSB)
0	0	0	0	0	0	0	0	0
10	0	0	0	0	1	0	1	0
16	0	0	0	1	0	0	0	0
50	0	0	1	1	0	0	1	0
100	0	1	1	0	0	1	0	0
128	1	0	0	0	0	0	0	0
150	1	0	0	1	0	1	1	0
228	1	1	1	0	0	1	1	0
255	1	1	1	1	1	1	1	1

Figure 3.13 Unsigned denary and binary numbers. Note positions of 'most significant' and 'least significant' bits (MSB and LSB, respectively)

We decide on the accuracy and resolution we need and select the number of bits accordingly.

How about positive and negative numbers?

Or more properly 'Does the conversion of negative numbers affect conversion?' So far, we have not said how our microprocessor will handle both positive and negative numbers. In fact, only denary numbers in the range 0–255 have been specifically mentioned.

We now need to see what can be done to process both positive and negative numbers.

Microprocessor manufacturers use a technique known as *two's complement* for handling positive and negative numbers but it has disadvantages as well as advantages.

With only eight bits available how are we to represent and recognise negative numbers?

This is done by using a *sign bit*. In fact, we use the *most significant bit* (MSB) to represent a negative sign as shown in Figure 3.14.

Does this mean, however, that we now have the nonsense of having both plus and minus zero? It certainly looks like it! But, in practice this is not a worry for us because in two's complement the fourth pattern shown in Figure 3.14 means not just that the number is negative but that it is in fact −128.

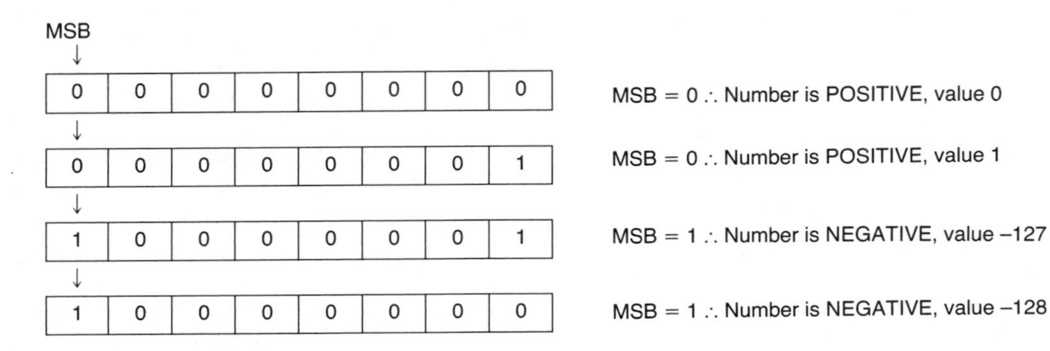

Figure 3.14 Effect of MSB on the sign of a digital number

Not only is the MSB acting as a sign bit, 0 for positive, 1 for negative, but its position gives it a value, i.e. when MSB = 0, the sign is positive, value = 0; when MSB = 1, the sign is negative, value = −128.

Let us look at Figure 3.15, a revised version of Figure 3.13, which now shows positive and negative numbers.

This is the beauty of the two's complement method; if the MSB = 0, then all the numbers are positive, in the range 0 to 127.

If MSB = 1, then all the numbers are negative, in the range of 128 to −1.

Does this mean, however, that we have lost the capability to represent numbers between 127 and 255? Yes, the main problem is that by allowing eight bits to represent both positive and negative numbers, the number value has been halved.

Denary number	Binary number -2^7	2^6	2^5	2^4	2^3	2^2	2^1	2^0
	−128	64	32	16	8	4	2	1
0 →	0	0	0	0	0	0	0	0
+ 1 →	0	0	0	0	0	0	0	1
+ 127 →	0	1	1	1	1	1	1	1
− 128 →	1	0	0	0	0	0	0	0
− 127 →	1	0	0	0	0	0	0	1
− 1 →	1	1	1	1	1	1	1	1

Figure 3.15 Positive and negative binary numbers using two's complement and their denary equivalents using eight bits

Note that in Figure 3.15 we get the idea that:

$$-127 = -128 + 1$$
$$-1 = -128 + 127$$

which is perfectly acceptable.

Figure 3.16 below shows that what has hap-

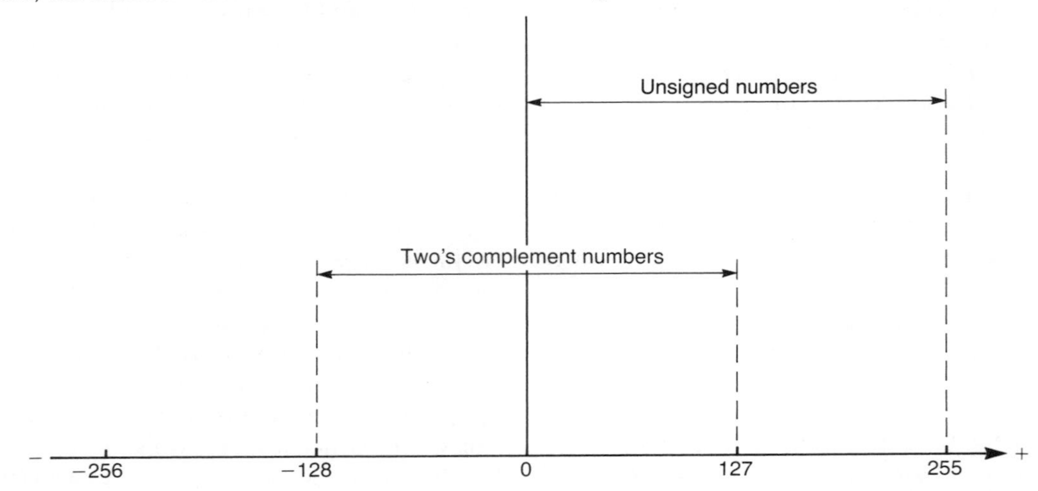

Figure 3.16 Number ranges for unsigned and two's complement numbers using eight bits

pened is a *shift* in the number range.

Number range
This shows that the advantage of using eight bits to represent positive and negative numbers is offset by the disadvantage of halving the size of numbers that can be processed,

i.e. 'positive or unsigned numbers' range 0 – 255
Positive and negative numbers, range −128 – +127

The analogue-to-digital converter (ADC) is usually an 8-bit device, manufactured in integrated circuit or 'chip' form, and is either organised to handle positive numbers or two's complement numbers. This usually has the effect of allowing the ADC to handle voltages in one of the two following ranges:

$$0 - 5\ \text{V}$$
$$-2.5 - +2.5\ \text{V}$$

How does the presence of electronic noise affect the process?
Noise can arise from a variety of sources: mains-borne noise, transient noise, pick-up hum, noisy components, etc. It is a real fact of engineering life that it is the noise present which affects the overall performance of the system.

If observed on a CRO, noise appears as fast spikes or pulses on signal lines and the tolerance of a digital system to them is usually not very great. The resolution of the system is influenced by the number of bits available in the ADC process. There is no point in using more than a certain number of bits if the amplitude of the noise present in the system is larger than the resolution itself. Thus the noise present will influence the choice of number of bits to be used and so affect the ADC performance.

Compromise has been made by the manufacturers of 8-bit processors to give an optimum performance of their devices against the contaminating effects of noise. They could improve resolution by providing more bits but each improvement in resolution is often counteracted by the noise levels causing errors to occur in the ADC process. In the end, it is control of noise that will lead to accurate, fast, reliable conversions.

CONVERSION TECHNIQUES

A variety of methods have been developed over the years. As processing power has developed and become faster, conversion time has put pressure on ADC designers also to make their products faster. One of the earliest methods used was called the single ramp method. This was used in early digital voltmeters. It was slow and cumbersome and difficult to set up and calibrate, which means adjusting the performance so that a reliable relationship can be achieved between the value of the input analogue voltage to the ADC converter and the resulting output digital signal itself. It was succeeded by the dual ramp method which was a great improvement, but was still slow.

A major step forward was made by developing the successive approximation method which not only was a lot faster than previous methods but had the advantage that the conversion time was the same irrespective of whether the input signal was large or small.

One of the fastest methods today is the flash technique but it is also, as you might expect, one of the most expensive.

Summary

In this chapter the major features of analogue and digital signals have been described and compared. The relatively slowly changing properties of signals, leading to the concepts of *amplitude*, *frequency*, *phase* and *periodic time* have been developed, together with the view that the real world in which we operate is essentially a continuous or analogue one.

The digital world, however, is of discrete states with gaps between the states which are unavailable to us to use. However, the pulses of the digital world are ideally of the same amplitude, frequency, periodic time, etc. and it is only their presence or absence which concerns us.

The concept of positive and negative binary number was then developed by using the *two's complement* method of operation. This allows a range of signals to be processed but at the price of halving the magnitude of the signals applied.

Suggested Assignment Work

1 Using 4-bit, 8-bit, 12-bit and 16-bit data bus structures calculate

(a) the range of unsigned numbers available

(b) the range of two's complement numbers available.

2 In one of the diagrams a device called a 'sample and hold' circuit is used. Explain why such a device is necessary.

3 Investigate practically the operation of an 8-bit DAC.

4 Using reference data sheets find out how ADC devices are packaged in integrated circuit form and their range of operation.

FOUR

Integrated circuit systems

OBJECTIVES

At the end of this unit all students should be able to:

Explain briefly the structure and interconnection of components on an integrated circuit and identify the scale of the circuit in terms of the chip device count.

So far all the systems that have been considered have been treated as boxes in a block diagram or symbols on flow charts which is a method of representing how software can react or work with the hardware to produce a desired result. This can be as straightforward as providing 100 watts of sound from each channel of a modern compact disc player to the very rapid computing and control systems in an active suspension unit in a modern Grand Prix racing car.

What has not been mentioned at all up to now is anything about the microelectronic systems that make this technology possible and, in particular, the integrated circuit systems which have led to development of the microprocessor system as a computing and control element. It has been the immense reduction in scale of microelectronics that has brought computing power down to components the size of matchboxes and even smaller.

Some of the processes that are used by the manufacturer of integrated circuits will be described, in order to understand how the chip is built up from electronic material

The manufacturer has to work to very critical limits of position, temperature, time and purity of material to obtain satisfactory performance from his integrated circuits. He has to develop as many devices in as few stages as possible to achieve reliability and consistency of operation over the large numbers of chips actually produced.

The manufacturer can produce, by using similar processes, a range of values of resistance and capacitance but *not* inductance. He can also produce diodes and transistors and connect all the devices in such a way in his design to achieve the best performance from the smallest area of material available to him.

While studying other subjects you *may* have come across the terms pn junction, diode, pnp, npn, transistor, MOS, FET, MOSFET, CMOS, etc. The types of devices available get more numerous by the day and trying to understand and learn about them becomes increasingly difficult.

However, as far as this book and subject are concerned, you will only need to know a little about the devices and what they do, so we will look at an npn transistor, at its structure and how it can be made by integrated circuit methods.

Figure 4.1 shows two components, a resistor R and an npn transistor. The resistor obeys Ohm's law which means that as the current passes through it, the potential difference (p.d.), a *voltage*, developed across it depends on the value of current and the *resistance value* associated with it, to give the version of the law:

$$V = I \times R$$

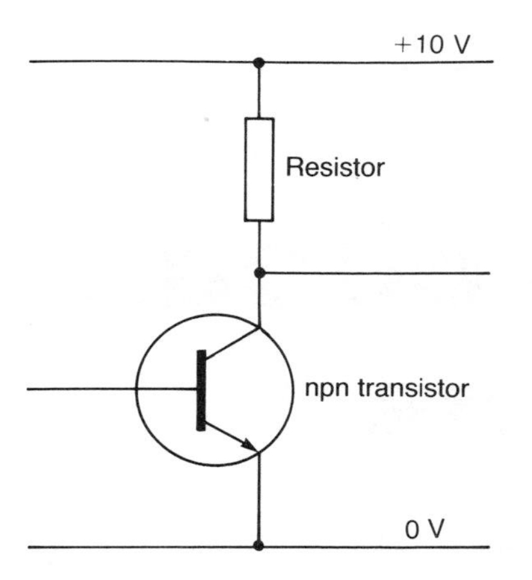

Figure 4.1 Simplified circuit of a transistor amplifier

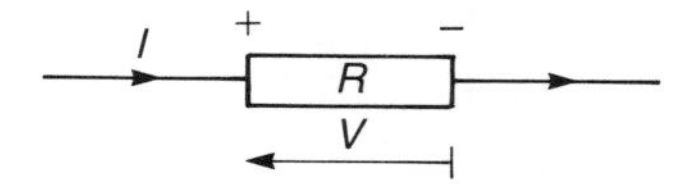

Figure 4.2 Effect of current flowing in a resistor and the production of a potential difference by Ohm's law

So if the current I doubles in value the value of the p.d. also doubles and we call R the resistance *constant*, and give it a value in ohms (Ω). You will use resistance in all sorts of ways in electronics and microelectronics but this law is fundamental and always works.

The other component is much more complicated. It is an npn transistor but *what* does that mean and *what* does it do?

The word *transistor* is made up from two words: *trans*fer-res*istor*. It acts as an *amplifier*, which makes very small signals very much larger by electronic means. For example, a compact disc player reads the disc by using a laser beam and makes the signals larger by using transistors to give us sufficient audio power in our loudspeakers for us to hear them.

What the transistor does is to make in one component two resistors, one very much larger than the other, in such a way that the current passing does not change very much, and so will produce a larger voltage, as shown in Figure 4.3.

What could happen is that V_{IN} is the very small signal voltage produced when the laser beam

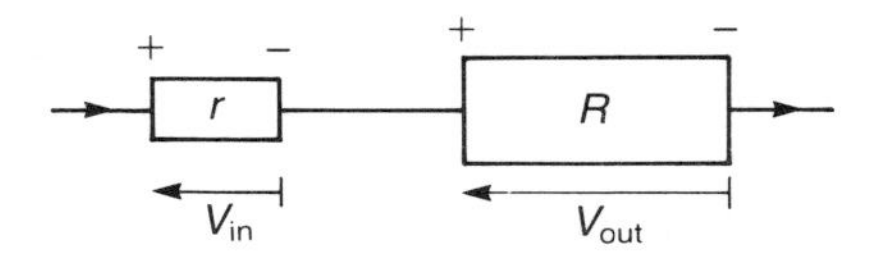

Figure 4.3 TRANSfer resISTOR

reads a track on the compact disc. The transistor, being made up of these two resistors of widely differing values r and R, allows the current produced when V_{in} is applied to r to flow across to R unchanged.

This current, flowing in this much larger resistor R, then develops a much larger voltage which we call V_{out}.

This greatly simplified explanation shows what the transistor does but it is only by careful design and manufacture that it works at all!

As the transistor is the vital part of any microelectronic circuit we will examine it a bit further.

The symbol shown in Figure 4.4 is that of an npn bipolar transistor. You will learn more about this device in electronics but a simplified explanation will be given here to help you try to understand the action of it.

Bipolar means that the internal workings depend on *two* types of current carrier, *electrons* and *holes*. Both exist in semiconductor material, such as silicon and germanium but the manufacturer can, by complicated and carefully controlled methods, *vary* the amounts of these electrons and holes. These amounts are called *doping levels* because small amounts of different substances are introduced into the basic semiconductor material at high temperatures. This doping will

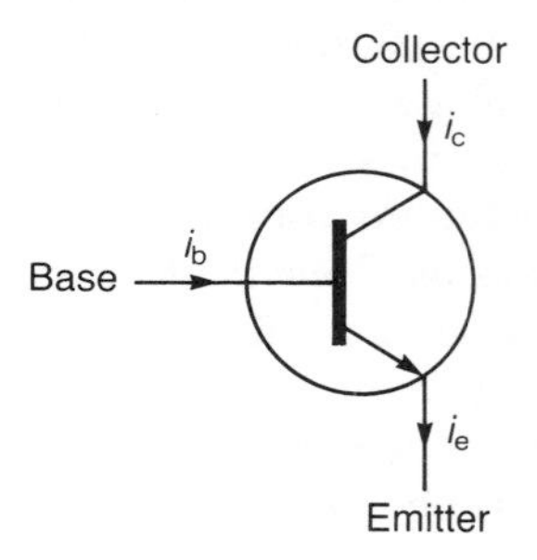

Figure 4.4 Names of the three terminals for an npn transistor and the directions of current when flowing in it

change both the physical properties and more importantly the electrical properties of the material.

The *two* types of current carriers that are produced by these methods are called *electrons* and *holes*.

As you will have learnt from your science teachers, all matter is made up from very small atomic particles called *protons*, *neutrons* and *electrons*. All natural materials are electrically neutral which means that any number of positive charges is exactly balanced by the same number of negative charges.

Protons have *positive* charges and are relatively *heavy*.
Neutrons have *no* charge and are relatively *heavy*.
Electrons have *negative* charges and are relatively *light*.

The *nucleus* of an element, such as carbon, germanium or silicon, is made up from protons and neutrons and ends up with a number of positive charges. Around the nucleus, in well defined paths, electrons circle in orbits in ways comparable with the way the planets in our solar system orbit round the sun. The number of electrons is such that the number of negative charges exactly balances the number of positive charges.

VALENCY

Silicon and germanium are known as *semiconductors*. For many years nobody would use them either as conductors of electricity because their resistivity was too high or as insulators because their resistivity was not high enough! As a result they were called semiconductors and were a curiosity.

One common feature about these semiconductors is that they are part of a group of elements in the Periodic Table of chemical elements with *four* valency electrons.

What does this mean? Quite simply, valency tells us the *number* of electrons in the orbit or shell of electrons *outermost* from the centre or nucleus of the atom itself, as shown in Figure 4.5 below.

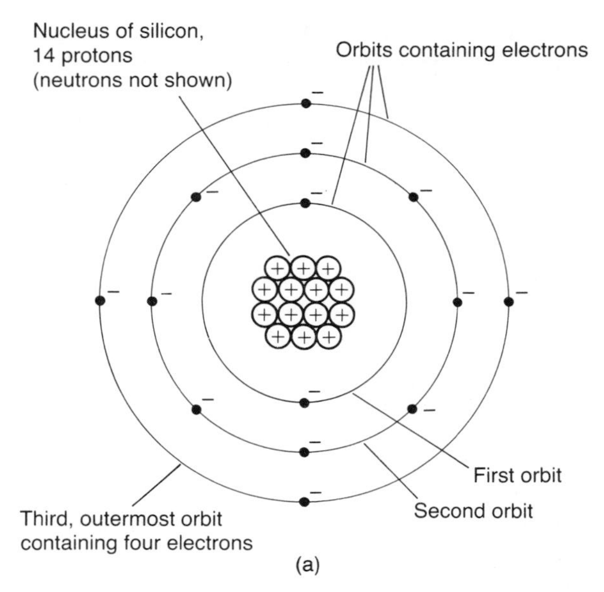

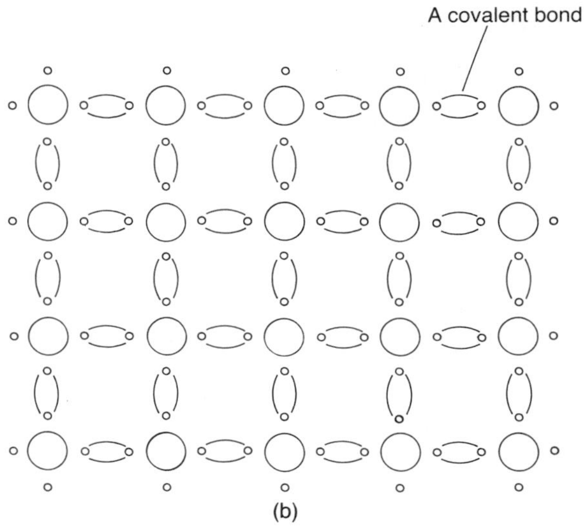

Figure 4.5 (a) Model of silicon atom and (b) simplified model of the silicon crystal lattice

Now, without going into a lot of unnecessary theory, what the semiconductor manufacturer can do is to change the electrical properties of the silicon by *adding* or *removing* electrons from the outermost orbits or *valency* levels. How the manufacturer does this is by replacing some of the silicon atoms in the crystal with atoms of another element. This is done in very small, carefully controlled quantities and is called *doping*. Such elements as boron, gallium and indium will produce one type of effect whereas antimony, arsenic and phosphorus will produce another. The semiconductor manufacturer can use these effects when making transistors.

If you look at Figure 4.5(b) you will see a diagram of an array which represents a pure silicon crystal. Each atom is bound tightly to its neighbours by bonds called *covalent* bonds. These are bonds between an electron in the outermost orbit of one atom, called a *valency* electron and a partner which is a corresponding valency electron belonging to a neighbouring atom.

To distinguish between the pure silicon and the silicon crystal into which these other elements are introduced, we use terms such as *impurity* atom, *intrinsic*, *extrinsic*, *donor* and *acceptor*. These will be explained as they are met, but for the moment consider the 'pure' silicon crystal as 'intrinsically' pure (hence intrinsic material) and the crystal to which the small amounts of other elements are added, i.e. which is doped, as 'extrinsically' pure (or extrinsic material). These other elements we call 'impurities', *not* because they are impure themselves but to show they are *different* elements from the pure silicon.

This raises a number of questions.

(1) *Does this process change silicon into some other element?*
No, because only very small numbers of *individual* atoms in the crystal are affected by this process. Typically only one atom in many millions is affected, so the material is still, basically, silicon.

(2) *Does this process change the neutrality of silicon because of adding or removing electrons?*
Again no, because the way in which these changes are made to happen by the semiconductor manufacturer is by *adding* small amounts of *impurity* atoms which are:

Trivalent – valency of three electrons, or
Pentavalent – valency of five electrons

Do not forget, however, these impurities are themselves electrically neutral so there will be *no* change in the electrical neutrality of the doped silicon.

(3) *This process must change the structure in some way but how?*
Yes, the structure is changed, but in very carefully controlled ways. Figure 4.7 further on attempts to show the effects of these changes. To help simplify this, a single silicon atom is shown in Figure 4.6.

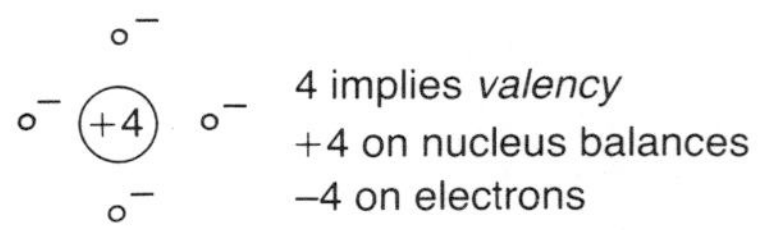

o⁻
o⁻ (+4) o⁻
o⁻

4 implies *valency*
+4 on nucleus balances
–4 on electrons

Figure 4.6 Simplified model of a silicon atom

Each nucleus is *bound* into the material by *covalent bonds* which give the silicon its mechanical strength and also its electrical properties of high resistivity. The covalent bonds are pairs of electrons, each one coming from one of a pair of neighbouring atoms. As semiconductors have four valency electrons there are no spare electrons that are free to move about in the material, which causes its resistivity to have a high value.

This is, simply, a reason why semiconductors act as they do, having a resistance which is too large for conduction purposes, as in wires, but too small for them to be insulators.

(4) *How do these trivalent or pentavalent impurities affect the silicon?*
By removing *some* of the silicon atoms and replacing them with the impurity, we create extrinsic semiconductor material. The electrical properties of this new material are just right for making *diodes*, *transistors*, etc.

Figure 4.7 shows the effect. The impurity atom of valency five (pentavalent) has five valency electrons. Four are involved in forming covalent bonds with neighbouring atoms, leaving one electron free or spare and this can move around quite freely. Because this free electron is a negative charge, this is called *n–type* material.

If a trivalent impurity is used, a different effect is created. Figure 4.8 opposite shows this effect.

The impurity atom of valency three (trivalent), has three valency electrons. All three are involved in forming covalent bonds with their neighbours. One of the neighbouring atoms has one valency electron which is prevented from forming a cova-

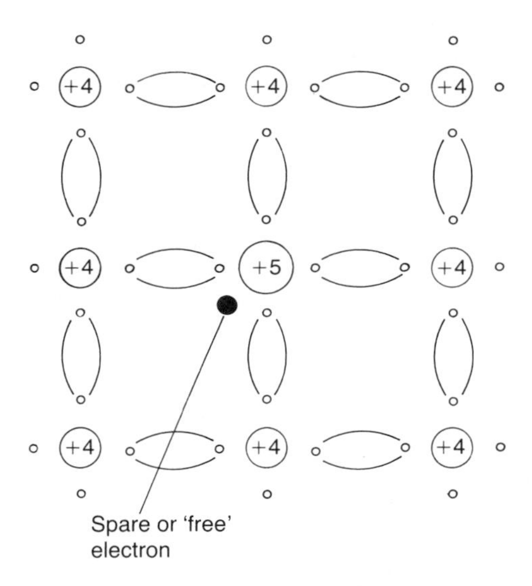

Figure 4.7 Crystal lattice of extrinsic, n-type semiconductor material

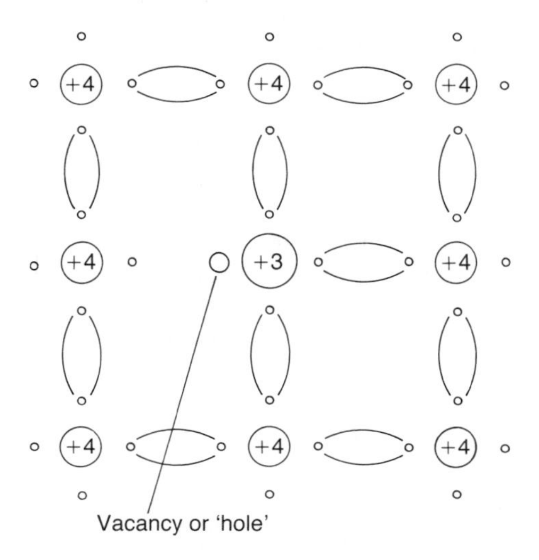

Figure 4.8 Crystal lattice of extrinsic, p-type semiconductor material

lent bond because of the hole in the outermost orbit of the trivalent impurity atom. This upsets the equilibrium of the system and the electron, looking to form a covalent bond, is said to be 'unsatisfied'. As will be explained next, this hole appears to act just like a positive charge and the material is called *p-type*.

An extrinsic semiconductor material whether of p- or n-type, will have a conductivity which is higher than for the pure intrinsic material, owing to the 'doping' (the addition of the impurity elements). The resulting free electrons, which are negative charges, and holes, which act like positive charges, can help current flow more easily through the extrinsic material and, as a result, the electrical performance is changed.

(5) *Can this hole move around at all?*
Yes, the neighbouring silicon atom, which has an 'unsatisfied' electron or incomplete covalent bond tries to obtain or 'grab' an electron from a neighbour. At normal temperatures silicon atoms have enough energy to do this at certain instants in time. Covalent bonds break and reform elsewhere. The 'hole' appears to move around the extrinsic semiconductor and also, more importantly for us, appears to act just like a positive charge!

Let us look again at the npn transistor that was shown in Figure 4.4. This can now be redrawn in terms of the types of semiconductor material, n and p, as shown in Figure 4.9.

The arrow, or rather the *direction* of the arrow tells us the type of transistor. Figure 4.10 shows another common example, a pnp transistor, usually made from another semiconductor material, germanium.

(6) *How does the manufacturer use these transistors to make integrated circuits?*
The manufacturer has a number of problems to overcome. Firstly, the semiconductor material, the silicon or germanium, must be made as pure as possible because the electrical properties of the transistors that are produced will depend on this.

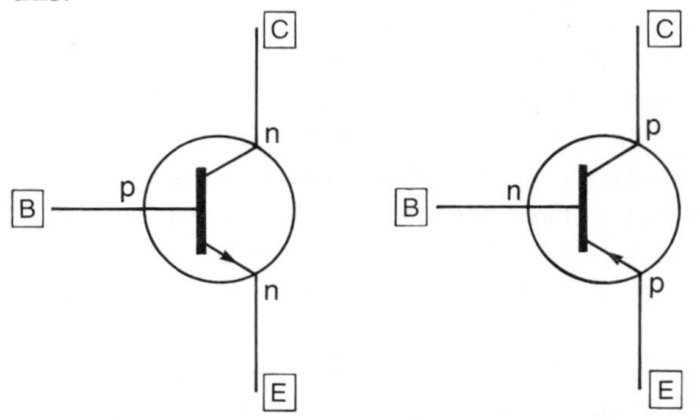

Figure 4.9 Symbol for an npn transistor

Figure 4.10 Symbol for a pnp transistor

What is produced as the basic component is a very large crystal of silicon or germanium, which looks like a metal rod, about 5 cm in diameter and about 50 cm long. It should be as pure as possible. This is then sliced into very thin discs called *wafers*.

These wafers are then treated with very small amounts of impurity atoms, either trivalent or pentavalent in nature, in very carefully controlled amounts to change their electrical behaviour into either p-type or n-type material. This is done by heating the wafers up in a furnace and then allowing a gas, which carries the impurity atoms in a gaseous form, over the wafer. The impurity atoms diffuse into the pure semiconductor and produce the doping effect which converts the pure material into p-type or n-type. This changes the intrinsic material into the extrinsic form that was mentioned earlier on.

The various stages are shown in Figures 4.11 and 4.12.

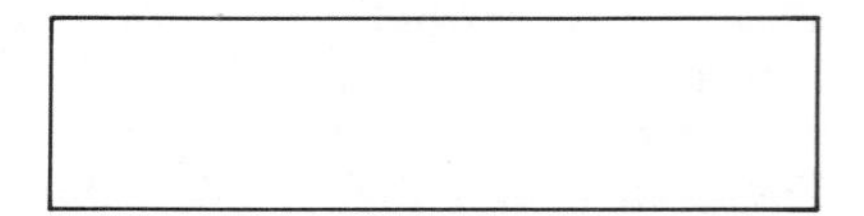

Figure 4.11 Wafer of pure silicon

The wafer is covered in a thin layer of 'photoresist', a material which is sensitive to light and can be processed, just like a film in a camera. A mask is made which when placed on top of the photoresist layer will shield it from the ultraviolet light the surface of the wafer is exposed to. As Figure 4.12 shows the ultraviolet light hardens the exposed areas but leaves the area under the mask soft and able to be etched away by chemical action. This etching opens up areas of the silicon for diffusion of the impurity to take place, but will protect the rest of the surface. As a result, the diffusion takes place in carefully controlled areas.

Figure 4.13 shows what happens when the impurity is diffused into the semiconductor itself. An n-type layer is formed which is very thin, or shallow, on the surface. By repeating the whole process, a p-type layer can be formed in the n-type region, leading to the development of a pn junction, as shown in Figure 4.14 below. This pn

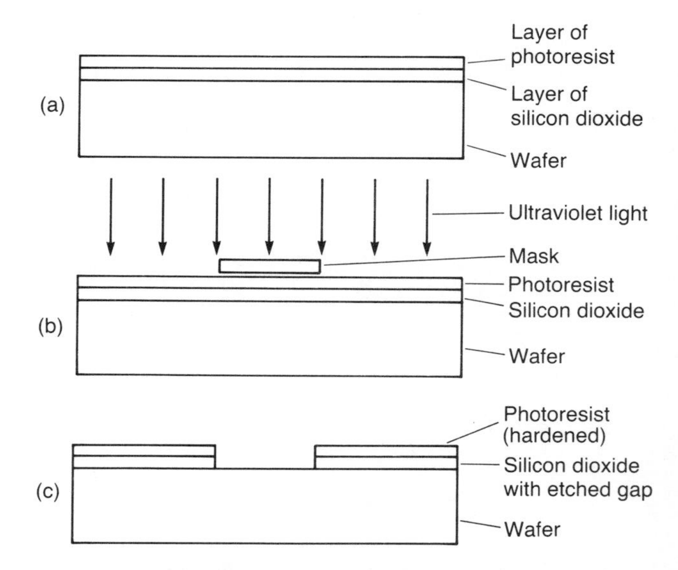

Figure 4.12 Stages of exposing a masked area of 'photoresist' to ultraviolet light. Subsequent chemical processing, like developing, causes a hole to be made through to the surface of the silicon wafer

junction is what the manufacturer wishes to make. It is an essential feature of transistors.

This explanation of the process is greatly simplified. It does not show how one transistor may be separated from another, how physical connections are made or other interesting constructional details. Various other activities take place to improve the electrical properties of the devices produced.

A diagram of an npn transistor is shown opposite in Figure 4.15.

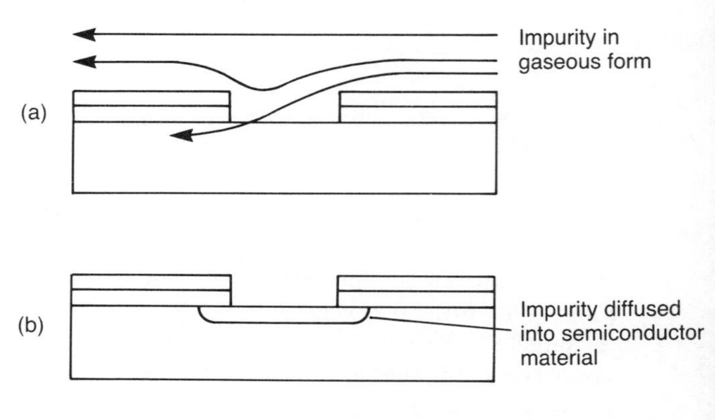

Figure 4.13 Impurity diffused into semiconductor material

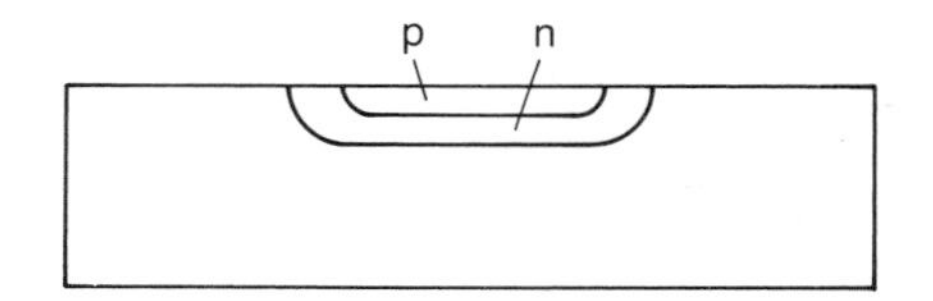

Figure 4.14 A pn junction formed in the surface of a wafer

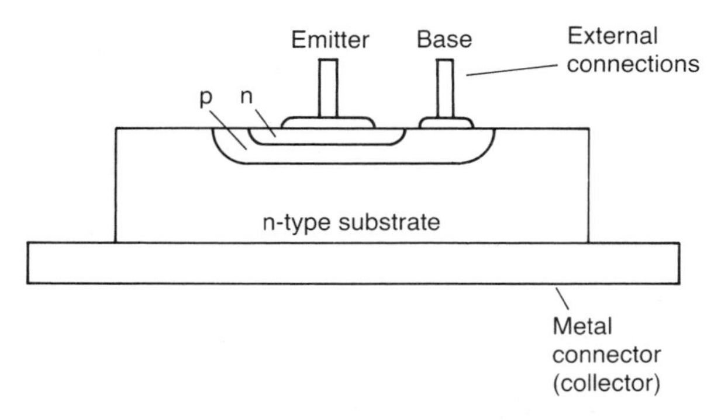

Figure 4.15 A completed npn transistor, mounted on a metal plate which acts as COLLECTOR. The EMITTER and BASE are connected to their respective regions with great accuracy

(7) *What are the impurities that are used?*
The trivalent impurities that create holes and hence produce p-type material include boron, gallium and indium. These are called *acceptor* impurities because they create vacancies ready to accept an electron.

The pentavalent impurities that donate free electrons and hence n-type material include antimony, phosphorus and arsenic. These are called *donor* impurities because they 'donate' a free electron.

(8) *How can this process be adapted to produce resistors and capacitors?*
Capacitance is always formed at a pn junction. By varying the geometry, a pn junction can be made of the required dimension to produce capacitor values of picofarads or nanofarads, which are often acceptable and useful for circuit designers.

Resistors are made by using the semiconductor property itself. By forming layers of extrinsic material in reasonably long diffusions, resistance values of up to 20 kΩ can be generated quite easily.

COMPLEX INTEGRATED CIRCUITS

Once manufacturers succeeded in making pnp and npn transistors, resistors and capacitors, they had the ingredients to make most electronic circuits in an integrated block. All that was needed was a way of separating the components whilst allowing them to be connected electrically.

This was done by deliberately doping the areas between components so that they were electrically isolated from each other and then depositing a metalisation (a thin layer of metal that can physically make the necessary electrical connection between components) over the surface to provide the necessary connections.

Figure 4.16 shows the principle of this operation.

The manufacturer can now make the integrated circuits as small or as large as required. Early integrated circuits, such as the *TTL (Transistor–Transistor Logic)* 74 series, which is a range of digital integrated circuit chips, has some members which have only a few resistors and transistors; others may have many hundreds or even thousands of components built up together to form large integrated circuit systems.

SCALE OF INTEGRATION

Many data books, reference books and manufacturers' details mention terms such as SSI, MSI, LSI or VLSI. Now we will find out what these terms mean and, more importantly, discover how many devices each one could hold.

SSI – Small Scale Integration
This was a term to describe small circuits, which could contain up to 10 different transistors. The original 7400 digital logic integrated circuit developed by Texas Instruments was produced as an SSI chip.

MSI – Medium Scale Integration
This is a term used to describe circuits of up to 100 components. Texas Instruments went on to produce more complex digital logic integrated circuits in MSI.

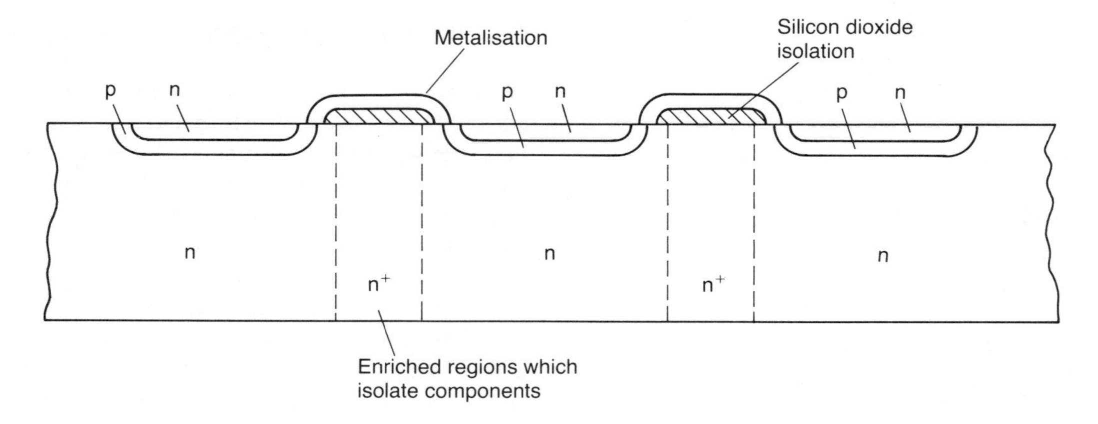

Figure 4.16 Enriched n-type regions acting as isolating parts between each of the required components

LSI – Large Scale Integration
A term used to describe circuit systems one order of magnitude higher, e.g. up to 1000 components. This would be used for some more complex controller chips, some peripheral driver chips, etc.

VLSI – Very Large Scale Integration
This describes systems with up to 10 000 components. It is used for most 8-bit and 16-bit microprocessor chips such as Motorola 6800 and 68000 or Intel 8085 and 8086 ranges of microprocessors.

This term does not only apply to the microprocessor but is also used to describe the other chips necessary to support the microprocessor such as PIA (Programmable Interface Adapter), VIA (Versatile Interface Adapter), ACIA (Asynchronous Communications Interface Adapter), PPI (Program Peripheral Interface), etc.

Nowadays, all manufacturers choose a technology, usually VLSI, or greater, to implement a range of integrated circuit functions they consider will answer a need in the market place, function well and reliably, be cost effective and make a product name and profit for the manufacturer concerned.

It is the application and use of VLSI as microprocessor and microcomputer chips that has revolutionised technology over the last few years. It has led to the ability to 'build in intelligence' to many devices and artefacts. This has been responsible for the tremendous rate of change in technology today.

SIZE OF INTEGRATION

Microelectronics is a word used to describe the very small structures on the surface of the silicon wafer. To give an indication of the size of these structures, modern VLSI chips of 10 000 transistors would be manufactured on a piece of silicon a few square millimetres in area! It is the subsequent packaging that makes the device appear larger. Even very large chips of 64 pins or more contain an active chip area not much bigger than the size suggested above.

Summary

In this chapter the concepts of semiconductor theory and its application to microelectronics have been developed.

The development of the transistor has been shown as a system, without the details of semiconductor physics. The chapter led on to the introduction of the two types of *transistor*, npn and pnp, with circuit diagrams.

Next, the terminology and processes of producing *n-type* and *p-type* semiconductor material from the intrinsically pure silicon or germanium was introduced. This was presented in a question and answer fashion to allow a gradual explanation of the various processes.

The methods of making an integrated circuit

from transistors, resistors and capacitors was then shown with brief explanations of the processes involved.

Finally the descriptions of *SSI*, *MSI*, *LSI* and *VLSI* were explained in terms of the component count on each particular type, together with a look at the size of a typical VLSI chip after manufacture.

Suggested Assignment Work

1 By reference to manufacturer's data sheets find out the meaning of the following terms:

TTL, CMOS and ECL.

2 Explain the main differences between these three types of operation and then compare all three in terms of power consumption, speed of operation and how they are affected by changes in temperature.

3 Investigate manufacturer's data sheets for the TTL devices and list by device number (e.g. 7400) and logic function (e.g. quad, two input NAND gate) for as many different types of gate as you can find.

4 Repeat exercise 3 for CMOS 4000 B series devices.

5 Investigate the data sheets for the Motorola 6800 and Intel 8085 microprocessors. Compare and contrast them in terms of:

data lines width
address lines width
addressing range
clock speed
type of technology of manufacturer.

FIVE

The microprocessor as a system

OBJECTIVES

At the end of this unit all students should be able, with the aid of a block diagram, to:

Explain the operational function of a microprocessor system comprising:
- *clock*
- *instruction register*
- *arithmetic and logic unit (ALU)*
- *memory (ROM and RAM)*
- *input/output devices*
- *bus lines*

This is a good point to re-examine the main components of any system. Figure 5.1 below is a diagram which was first shown in Chapter One.

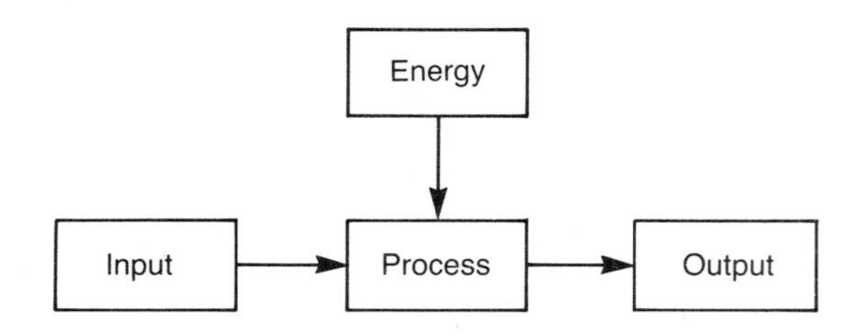

Figure 5.1 Block diagram of a system

The microprocessor, as a system, uses energy to process inputs before making them become outputs and is therefore no different from any other of the systems met so far, but there is *one* major difference between the microprocessor and all other systems. The microprocessor can be *programmed* to carry out a simple set of instructions and so is often described as possessing intelligence because it can make decisions, add-up, subtract, carry out logic functions. Later, another chapter will examine some of these functions that can be programmed into the microprocessor.

Before that can be attempted, however, the internal hardware of the microprocessor will be examined in general before an investigation of two of the most commonly found 8-bit microprocessors in use today, namely the Motorola 6802 and Intel 8085. Externally they appear to have similar specifications and performance, but internally they are quite different from each other.

The reason for choosing these two particular examples is quite simple. The two manufacturers, Motorola and Intel, command about 75 per cent of the world market for microprocessors.

As this book is concerned only with 8-bit devices it seemed appropriate to choose two devices, one from each manufacturer, that represented the development of 8-bit technology. There are other, very interesting, 8-bit processors but most of them in architectural terms (or how they were designed to be built!) are dependent on these two giants of the industry.

One of the first details to get to grips with in microprocessor technology is the nature of the 'buses'. Just as a bus carries people from one place to another for all sorts of reasons, going to work, to a football match, to the cinema, so also do the buses in a microprocessor based system transfer instructions, data and control signals all over the system.

There are three main buses in use to make the system work. They are:

address bus
data bus
control bus

Unlike motor buses which carry many people around at any one time, these buses carry single bits of instructions or data or control and are under the control of a master clock which times and controls the whole pace of operation. Also, microprocessor buses work in *parallel*, either 8 bits or 16 bits wide at a time.

DATA BUS

For both Motorola and Intel microprocessors, this is an 8-bit wide bi-directional data bus.

Figure 5.2 shows a microprocessor, symbol μP, to which are connected eight bits or eight lines of *two-way data bus*. They have to be two-way or bi-directional to allow data to flow both into and out from the microprocessor.

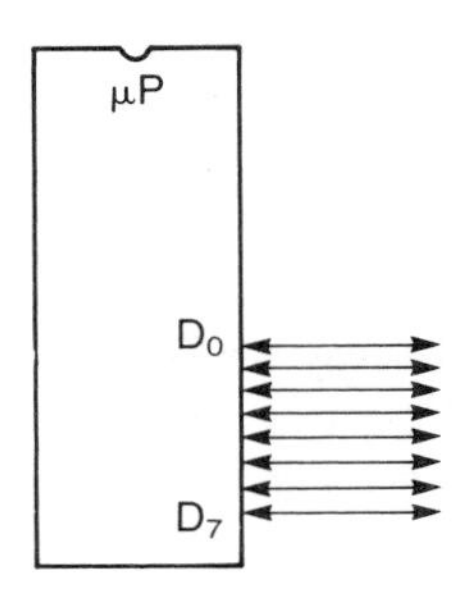

Figure 5.2 Schematic showing the bidirectional (2-way) data bus

If the data bus is eight bits wide, how many different digital combinations can be allowed?

In this case the answer is 256 (= 2^8) because there are 8 bits. Try it on a calculator and see for yourself.

This means there are 256 different digital combinations on a two-way 8-bit data bus.

Although there are eight data bits, they are described in a rather odd way. The smallest value binary digit (bit) is called the *least significant bit* (*LSB* for short) and has the value 2^0 which is 1.

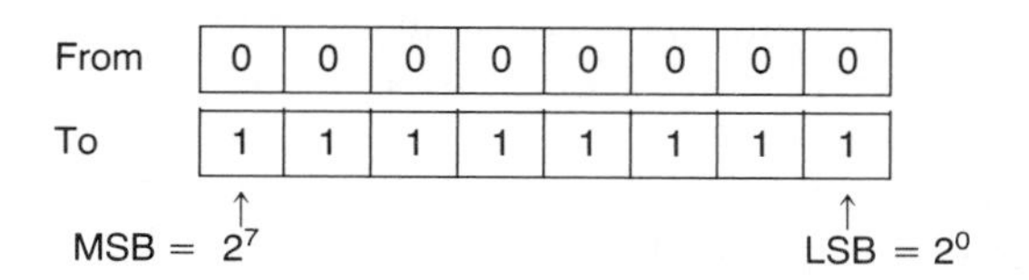

Figure 5.3 Smallest value and largest value bit patterns

i.e. a '1' in the LSB has value 1
a '0' in the LSB has value 0

This is repeated all the way through the sequence until the largest value bit is reached. This is called the *most significant bit* (*MSB* for short) and has the value

2^7 which is $2 \times 2 \times 2 \times 2 \times 2 \times 2 \times 2 = 128$

i.e. a '1' in the MSB has value 128
a '0' in the MSB has value 0

The wider meaning of this is that *each bit* has a value due to its *place* in the *data pattern*.

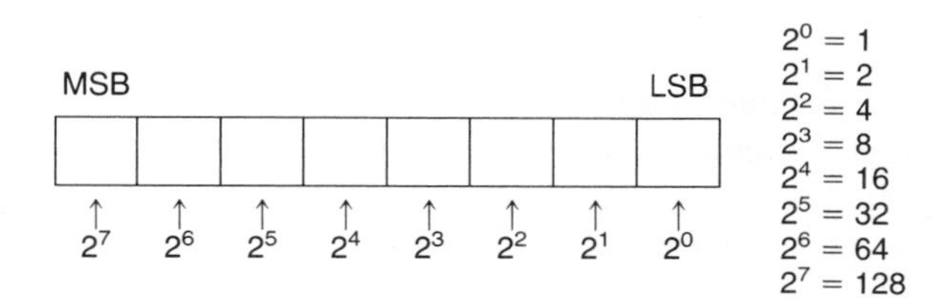

Figure 5.4 Values of bit positions in one byte

It follows then, for every cycle of the clock, that the data bus can have any one of 256 different combinations of bits. These range from the pattern shown in Figure 5.3 from *all* noughts, obviously zero, to *all* ones, not so obviously 255.

zero is the smallest number possible
255 is the largest number possible
} using 8 bits

Try it and see if it adds up to 255.

Figure 5.4 shows the value of each particular bit as a move is made up through the bits, from LSB to MSB.

This data pattern of eight bits is called a byte. This is a convenient way to describe a group of eight bits and, more particularly, the *value* of the pattern on the data bus of an 8-bit microprocessor.

Do not forget – this can be an instruction or data and the μP has to be able to decide, correctly, which it is. More about this later!

ADDRESS BUS

The microprocessor has to remain in control of the complete set of operations of the system which usually is made up as a *microcomputer*.

A microcomputer system is made up from the microprocessor itself, together with the other integrated circuits, or *chips*. These can be memory devices in which are stored the programs and data for the microprocessor itself plus the interface devices which are needed to allow the microcomputer system to communicate with the 'real world'. Remember from Chapter One that a *system* accepts *inputs* and after processing sends them back as *outputs*. There are interface devices where the inputs to and outputs from the system occur.

The microcomputer system, made up from these components, needs a certain amount of organisation and management. The way in which a microprocessor can rapidly attract the attention of these components is by using the *address bus*. This is a 16-bit wide collection of wires in parallel, starting at the microprocessor and going, in parallel, to all the components in the system, and lets the microprocessor control the system.

We will now look at the block diagram of a simple microcomputer system which is made up from the following chips:

- μP – microprocessor
- ROM – Read Only Memory
- RAM – Random Access Memory
- PIO – Programmable Input/Output chips

All these terms will be described fully a little later on pages 43–4. Figure 5.5 shows a system made up from these chips, all connected in parallel to the same data bus.

All four devices, μP, ROM, RAM and PIO, have the capability of 'talking' to the data bus, which means being able to put data onto the bus. Most of

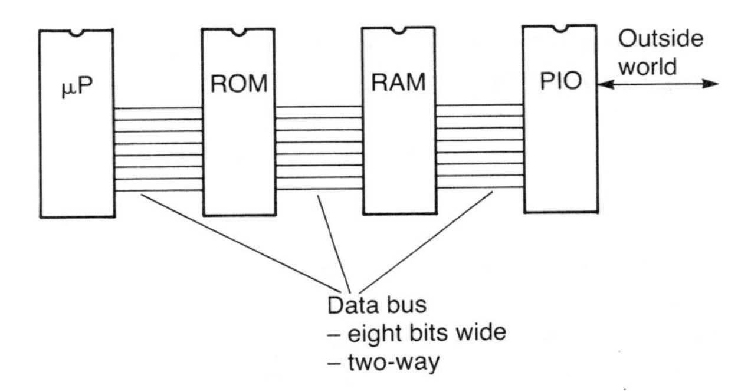

Figure 5.5 Schematic showing how the data bus connects all other devices on the system: in parallel, eight bits wide, 2-way

them, except the ROM, have a similar capability to 'listen' to the data bus which means being able to receive data from the bus.

These are often called 'write and read' operations. If all devices tried to do this at the same time, then confusion would happen! The jargon term for this is *bus contention*, when two devices try to get access to the data bus at the same time. It should *not* happen!

How does the microprocessor allow one device only to be connected to the data bus at any one time?

It does this by use of the address bus. For both microprocessors, this is a 16-bit-wide, unidirectional address bus. Figure 5.6 shows the address bus going out *from* the microprocessor.

It *has* to be one-way so that the μP can address a single, unique location or *address*.

It is a bit like a teacher asking a question and then naming a particular student to answer it. All

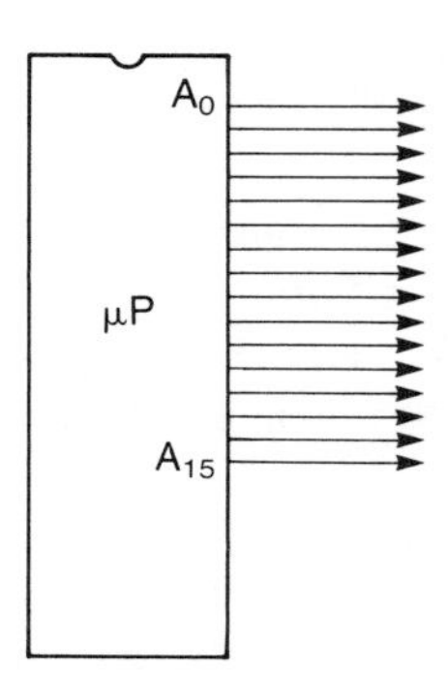

Figure 5.6 Schematic showing the unidirectional (1-way) address bus

the other students keep their heads down leaving the named student to answer the teacher!

So, the μP addresses a single location. This then sets up a situation which allows data transfer to take place.

If the address bus is 16 bits wide how many separate locations can be addressed?

In this case the answer is 65 536 = 2^{16} because there are 16 bits of address bus available. Once again, try it on a calculator to check it out for yourself.

This figure is often abbreviated to 64K.

The reason, once again using powers and indices, is that

$$2^{16} = 2^6 \times 2^{10} = 64 \times 1024$$

and *we* say that 1024 is equal to 1 K.

Note In the denary or decimal system, 1 K = 1000.

$$\text{Hence } 2^{16} = 64 \text{ K}$$

Just like the data bus, the address bus has to go everywhere in the system, so that each device knows where it is on the address bus, what its address actually is and how to recognise when the microprocessor wants it to 'talk' to it or otherwise transfer data, one way or the other. This is shown in Figure 5.7.

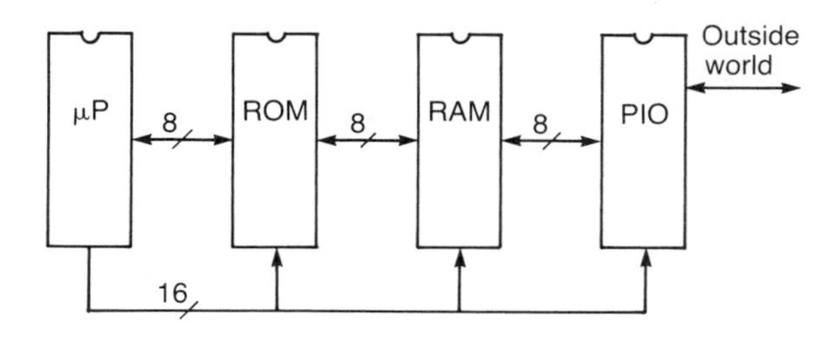

Figure 5.7 Simplified schematic of a microcomputer system

To save repeated drawing of eight lines or 16 lines at a time, the data and address buses are often simplified as shown in Figure 5.7. What *is* important is to recognise that:

the data bus is two-way and is *8 bits wide*
the address bus is one-way and is *16 bits wide.*

Semiconductor memory devices used in microcomputer systems usually come as one of two variants, *random access memory* (RAM) and *read only memory* (ROM).

RAM

This is memory in which the user can temporarily store programs and data. The user is also able to change or modify the contents of the program and its data used where necessary. Once the program is entered into RAM storage, the microprocessor is directed to obey the sequence of instructions and data stored there.

There are two different processes involved in this; the first is to 'enter' the programs and data into the store. This means that the user makes certain that the program of instructions to be obeyed and the data to be worked with are inserted into the correct places in the memory. These places have *addresses* just as houses do. This process is called *writing* to memory.

Once the program and its data have been entered into the store, the second process is to make the microprocessor go to those addresses where the program is stored and examine the instructions and data contained. This is called *reading* from memory.

Random access memory is thus capable of being *written* to or *read* from. Another description of RAM is to call it *read/write* memory.

As these memory elements receive electrical power to make them work, the shortest break in supply will cause the contents to become degraded, changed, corrupted or 'lost'. Just as the human brain will die if it fails to receive oxygen, RAM memory contents will fail if power is lost, even for a very short time. Such memory is called *volatile*.

ROM

This stands for *read only* memory. Unlike RAM, which the user can write to and read from, a ROM has its program and data contents permanently built into it, often by the manufacturer.

Because of this it can only be *read* by the user and not written to. An advantage is that a loss of power will have no effect, i.e. the contents will not become corrupted or 'lost'. Such memory is called *non-volatile*.

This is useful to the manufacturer as it allows for the permanent storage of such programs as the *operating system* supplied with the microcomputer. This allows the user to operate the system complete with keyboard, VDU screen, printer and disk drive easily and efficiently knowing that even if the power is lost the operating system will survive.

PIO (PROGRAMMABLE INPUT/OUTPUT DEVICES)

A microprocessor system is very little use on its own! It has to interface itself to the outside, 'real' world in which we humans live and work. If it is to be useful it has to work in 'real time' which is just another way of saying that it has to carry out its functions or jobs in a time scale which is acceptable to us as users.

The real world is predominantly analogue and relatively slow moving when compared with the speed of digital processing that is possible with a microprocessor. It should not be forgotten that a typical microprocessor works at a speed of 1 MHz, 1 million cycles per second.

The analogue-to-digital and digital-to-analogue converter chips that were mentioned in an earlier chapter are often of great use when interfacing a microprocessor to the outside world.

The PIO is a programmable device with a number of user-controlled functions. It usually has two *ports*, A and B, each of eight bits in parallel. Just like a port on the coast it is a place *from* which and *to* which items of use, such as programs and data can flow.

Each port is really an extention to the outside world of the 8-bit wide, parallel bidirectional data bus except that, under program control, the direction of each individual data line can be controlled. In other words on one port a number of different combinations are possible:

lines can be all input or all output
one line can be an input, and seven outputs
or any other combination to suit a particular application.

One of the major differences between Motorola and Intel is the way in which they handle the input/output process.

Having introduced the two main bus systems, the 8-bit, bidirectional data bus and the 16-bit unidirectional address bus, ROM, RAM and PIO, we now have the basis to form a viable microcomputer system of which the microprocessor is the heart.

There is one further bus system that needs to be mentioned now and that is the *control* bus. Unlike the two buses previously described, there is no standard size or direction for the control bus.

For the Motorola 6800 there are 11 control lines, eight as inputs and three as outputs.

For the Intel 8085 there are 14 control lines, five as inputs and nine as outputs.

As each manufacturer has developed their products, there has been a need to reduce the number of control lines to make room for other functions.

There are, however, a few control lines that need explaining:

R/W (READ/WRITE)

This is a control line issued by the microprocessor and is mainly used with random access memory (RAM), which can be written *to* as well as being read *from*. To tell the RAM what type of operation is needed the microprocessor places a logic 1 on this line to tell the RAM that the microprocessor wishes to *read from* that memory, whereas a logic 0 tells it that data is to be *written to* it.

TSC (TRI-STATE CONTROL)

The way in which the microprocessor prevents more than one chip fighting to gain access is by 'nominating' the particular chip it wishes to communicate with and to tell all the others that they should 'switch off' or take no further 'interest' until ordered to do so. This is done by use of the *tri-state logic* controllers.

A tri-state logic device is not much different from a binary logic device, except that it can also be switched into an 'off' condition. The three logic states are 0, 1 and *high impedance*. The first two are the usual logic levels of 0 V and 5 V, but the high impedance state has no voltage associated with it and will allow no current to flow when in that state.

The microprocessor uses this control line, which again is called different names by each manufacturer, to 'switch-off' or 'tri-state' as it is called, those chips that it does not want to communicate with.

INT (INTERRUPT)

This is a control wire that allows peripheral devices, such as a keyboard, disk drive, visual display unit, etc. to 'flag' to the microprocessor that somebody or something in the real world wishes to communicate with it.

The two manufacturers have different ways of handling interrupt but both recognise the importance of the technique and have arranged strategies for dealing with these events.

INTERNAL OPERATIONS OF A MICROPROCESSOR

All the preceding information refers to the *hardware* aspects of the system. This means the physical, tangible parts of the system that can be looked at, identified and examined by users of test instruments such as digital voltmeters (DVM), cathode-ray oscilloscopes (CRO) and logic analysers.

The microprocessor was described earlier as a programmable, clock-driven, digital integrated circuit. We now need to look inside the device to see how it works. Both Motorola and Intel use effective clock frequencies of around 1 MHz for their operations, but different instructions take different times to execute.

You will notice the introduction of two terms, *instructions* and *execute*. By 'execute' we do not mean kill or destroy, but 'carry out'.

In practice these two words describe the operation of any eight-bit microprocessor. It has a limited number of instructions it can recognise, and after recognising them it carries them out or executes them. The process is based on the 'fetch-execute' cycle.

What this means is that the microprocessor has to remain in control of the system at all times, except for some special operations, such as direct memory access (not described here) and carry out the instructions contained in either a user's program or part of the operating system contained in read only memory (ROM).

There are two internal parts of the microprocessor which are used to control this process. They are the *instruction register* and the *arithmetic and logic unit* (ALU). Between them they control the operation when it is carrying out a program.

The instruction set of any microprocessor is a list of instructions or *op-codes* as they are often called, and is presented in *hexadecimal* code, which is a compact way of representing binary code. This is explained further in Appendix C.

Typical instructions are simple ones such as:

LOAD ACCUMULATOR (ACC)	This means *loading* the accumulator with data from some other place. It is a process very like *writing* to a memory address location.
ADD ACCUMULATOR (ACC)	This means *adding* to the contents of the accumulator a number from some other place and changing the contents of the accumulator as a result.
SUB ACCUMULATOR	This means *subtracting* from the accumulator a number from some other place.
STORE ACCUMULATOR	This places a copy of the contents of the accumulator somewhere in memory storage.

Because each instruction is different in nature, they all take different times to complete or execute and so the microprocessor must be able to control the timing of each instruction, and do it in such a way that no corruption or distortion of the programmer's intentions happens. This is controlled by the *instruction register*, coupled with the *instruction decode register*. Register is a name used for part of the internal working of the microprocessor. This allows the microprocessor to recognise every possible instruction. The microprocessor is then able to decode and carry out these instructions properly.

But what is an *accumulator*? It is a word used for a general purpose register used by the microprocessor as part of its internal functions and also as a way of allowing the user outside the system to look at any result of an operation carried out by the system.

It is used in conjunction with the *arithmetic and logic unit* (ALU) which actually does all the number-crunching, Boolean logic operations and other decision making, etc. required of the microprocessor. The accumulator is the register which the microprocessor uses to communicate with the rest of the system to get the job done.

The instruction register is another register which is used to receive the instruction byte or op-code from off the data bus and to store it temporarily so that it can be decoded. From this decoding process the timing and control is set up to carry out the rest of the instruction. What happens is that, in the instruction decoding process, the instruction decoder compares the contents of the instruction register against its own contents. When it finds a match between them it then knows how to carry out the rest of the instruction because it has this vital information stored within its own structure.

Figure 5.8 shows an extract from Intel literature, showing the functional block diagram of the 8085A CPU (*Central Processing Unit*). The instruction register and instruction decoder are shown in the middle part of the figure and immediately to the left can be found the accumulator, arithmetic and logic unit (ALU) and *flags* register. This is a very useful register, whose eight bits can be individu-

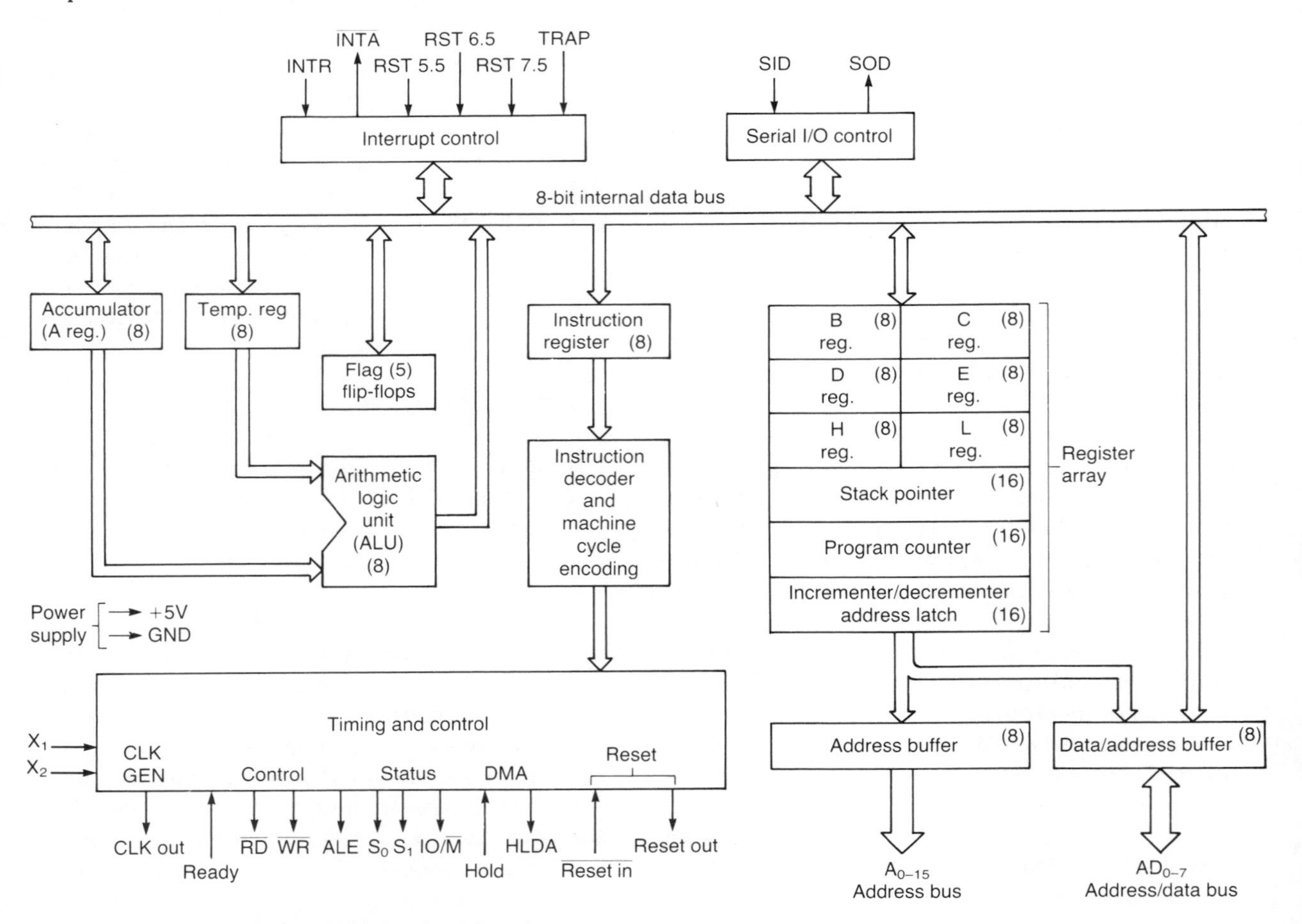

Figure 5.8 Intel 8085A CPU functional block diagram

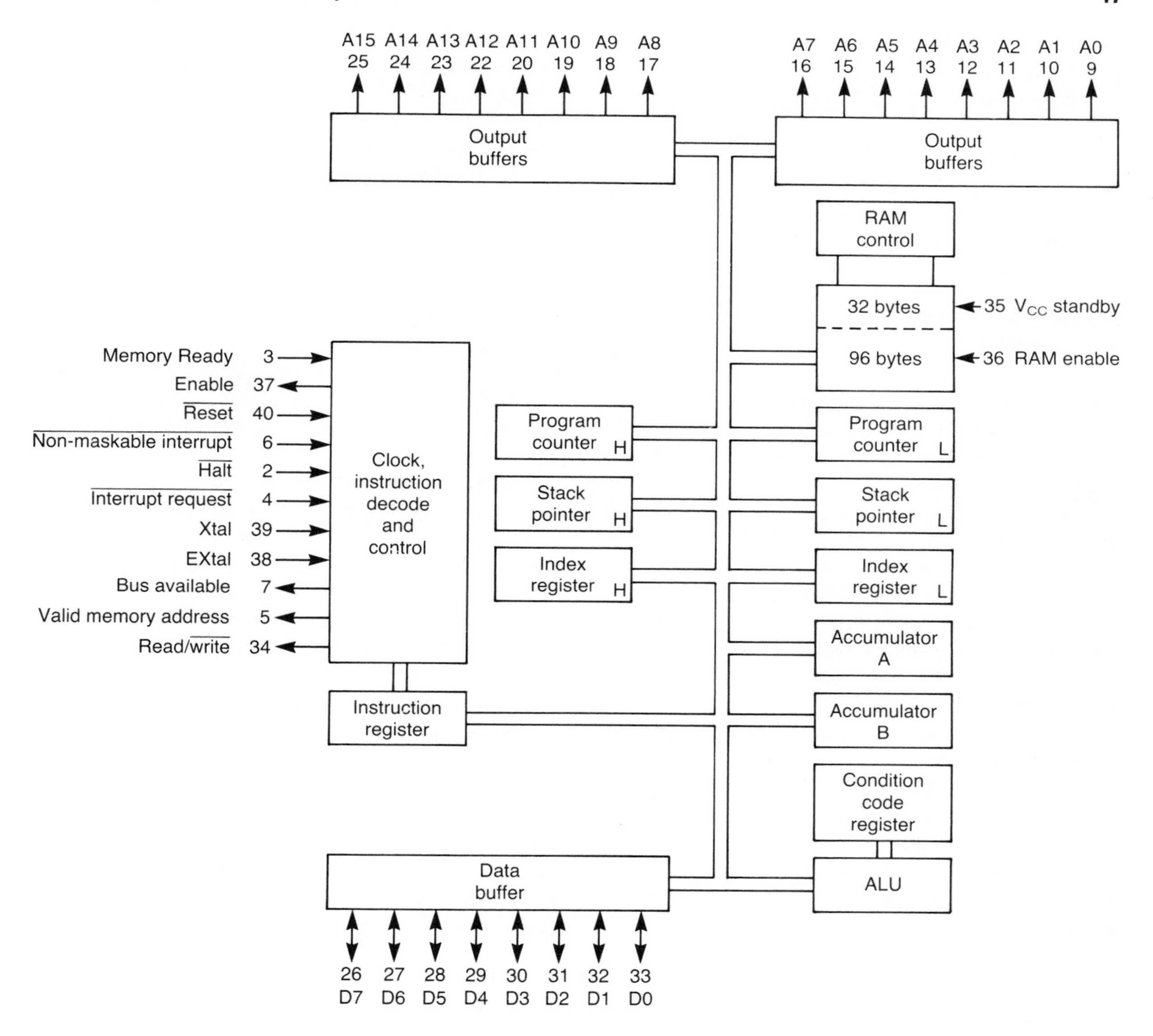

Figure 5.9 Motorola MC6802 expanded block diagram

ally set or reset, depending on certain key operations in the ALU. For example, if the result of an operation results in a zero value, then one particular bit of the flags is set to show that has happened. Other events flagged include parity, sign, carry and auxiliary carry.

Figure 5.9 shows an extract from a Motorola data sheet of the MC6802 expanded block diagram. This shows, amongst many other things, an instruction register, clock, instruction and decode control, *two* accumulators, an ALU and condition code register, which has a similar function to the flags of the Intel 8050A.

There are many other similarities between the two machines which will not be covered here.

Having dealt with the main functional details of the internal hardware of the microprocessor system, it is now time to consider the details of the fetch–execute cycle.

Figure 5.10 shows a simplified timing diagram

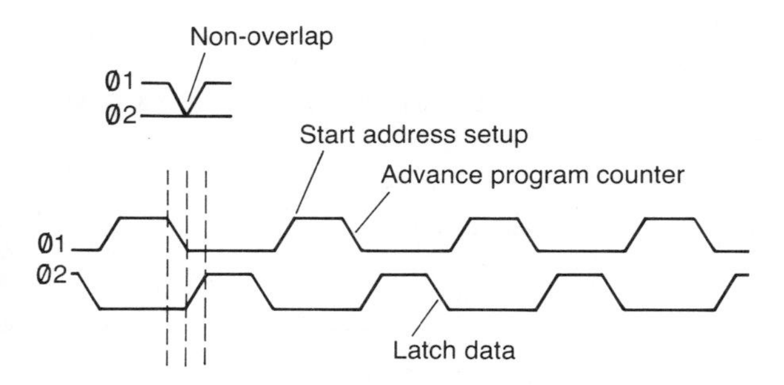

Figure 5.10 MPU clock waveform for the 6802 microprocessor

used by Motorola for their 6800 family series of microprocessors. It is not quite the same as the 6802 but will be adequate to describe the fetch–execute process.

During the rising edge of clock 1 (Ø1) an address, 16 bits wide, is placed on the address bus. Somewhere on the complete system that address will find a location in ROM, RAM or PIA, which is holding the contents which should be transferred to the instruction register, because it happens to be the instruction or op-code previously referred to.

On the falling edge of clock 1 (Ø1) the contents of the program counter are incremented by one so that, when it is called upon, it is already pointing at the next address to be put onto the address bus. The program counter is used by the system to point to the next location to be accessed and keeps track of the op-codes in any program the machine is running.

On the rising edge of clock 2 (Ø2) the contents of the original address, which should be an instruction, are placed onto the 8-bit data bus. On the falling edge of clock 2 (Ø2) this data (an instruction) is latched, or stored, in the microprocessor instruction register.

The instruction is decoded and this tells the control unit how many other addresses need to be accessed, or in other words how long the instruction is, before the instruction can be actually carried out or executed. At each stage, the program counter is advanced by one, as previously described, so that when it gets to the next relevant instruction, the MPU (microprocessor unit) is ready to repeat the whole process or another one, again.

Figure 5.11 shows how a similar process is carried out in the Intel 8085A and appears to be more complicated. The 'T states' are clock cycles and the M_1, M_2, M_3 and M_4 are called 'machine cycles'. In practice, the important point to note is that instructions are fetched from memory locations, decoded and then executed in much the same way as described previously.

The process is repeated again and again in a controlled manner until the entire program is carried out.

A note of warning, however; if *you*, the person writing the program, choose the wrong instruction the MPU has no way of knowing that and will try to carry out your instructions. When it gets lost or stops working because it cannot recognise or decode your program, it will be *your* fault!

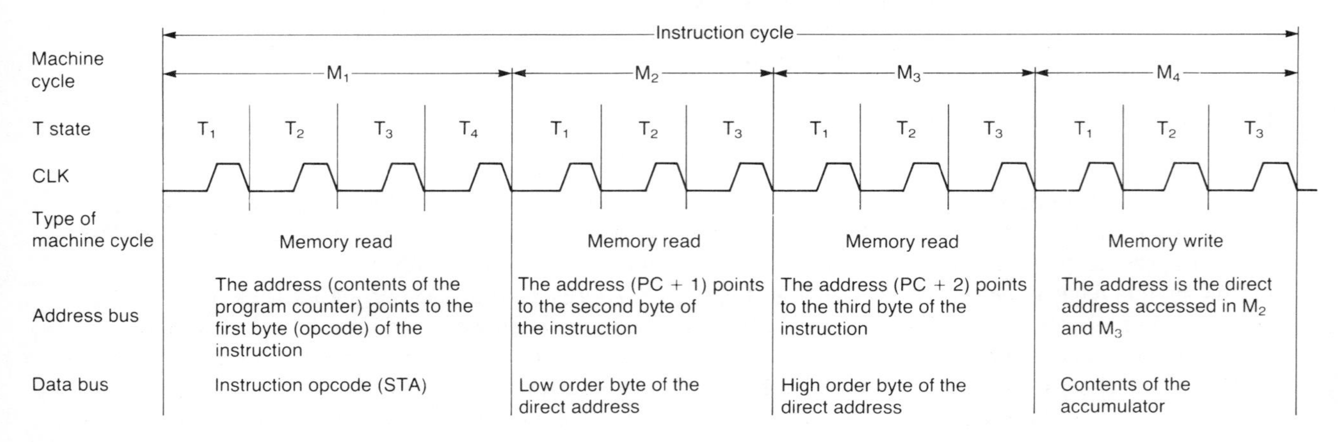

Figure 5.11 CPU timing for store accumulator direct (STA) instruction for the 8085 microprocessor

Summary

In this chapter, the microprocessor itself has been introduced as a system. Reference was made to two particularly successful 8-bit microprocessors: the Intel 8085A and the Motorola 6802, both with very similar specifications of data bus width, addressing range and other performance characteristics.

The terms of *data bus*, *address bus*, *read only memory (ROM)*, *random access memory (RAM)*, *programmable interface* and various necessary control lines have been introduced, along with an explanation of *tri-state devices*.

The internal operations of *instruction*, *fetch* and *execute* have been explained and developed so that a picture of a controlled, measured operation of the microprocessor system should now be understood. This has been helped by the use of diagrams issued by the two manufacturers involved.

Finally, in the warning contained at the end of the chapter the idea is also conveyed that the microprocessor is a very different integrated circuit from many earlier ones because it can be programmed to carry out a limited range of operations or instructions. It is this capability that has revolutionised electronics.

Suggested Assignment Work

1 Research data from other microprocessor manufacturers to discover the significant features of other 8-bit microprocessors, such as the Zilog Z80 and Mostek 6502.

2 Investigate Motorola data sheets to discover the major differences between the various interface chips such as:
peripheral interface adapter (PIA–MC6821)
asynchronous communications interface adapter (ACIA–MC6850)

3 Investigate Intel data sheets to repeat the discoveries of assignment 2 on their various interface chips such as:
universal peripheral interface – 8041
programmable communications interface – 8251A

4 In Figures 5.8 and 5.9 many additional internal registers and control devices are shown. Read the manufacturer's literature and compare how each manufacturer organises input/output operations.

SIX

Peripherals for microprocessor systems

OBJECTIVES

At the end of this unit all students should be able to:

Making reference to microprocessor operations, describe the basic functions of:

Visual display unit (VDU)
Teletype
Magnetic tape
Disk

In Chapter Five, the microprocessor was shown as a system which needs inputs and outputs. The internal structure and operation of the microprocessor relies on the data bus and address bus being connected to all other chips connected to the system including the peripheral interface chips. There are many types of these chips, each one being designed to carry out a particular function or to interface to a particular type of peripheral equipment.

In practice, it is the interfacing of the microprocessor system to the outside world, or real world, that often causes the microprocessor engineer the biggest problems or major challenges. Because the microcomputer is so much faster in operation than the human user the efficiency may be increased by allowing the system to handle a wide number of peripherals at any one time. To ensure that bus contention does not occur, a variety of methods are used but, in particular, it is the interrupt handling capability of the microprocessor itself that controls the access of the various peripheral devices to the microprocessor system itself.

It should be remembered that the interface chips which were briefly described earlier were said to interface the system to the outside world. In this chapter, the role of the interface will not only be examined briefly as a device in its own right but also as a buffer for the various peripheral devices that humans use, such as VDUs, teletypes and disc or tape back up storage.

PROGRAMMABLE PERIPHERAL INTERFACE (PPI)

The block diagram of this is shown in Figure 6.1.

Each port contains:
an 8-bit data direction register
8-bit data register (input or output)
8-bit control register.

By selecting the control register with a particular 8-bit code, the data direction register may be accessed. Each bit in this register controls the direction of data flow on the particular line of the

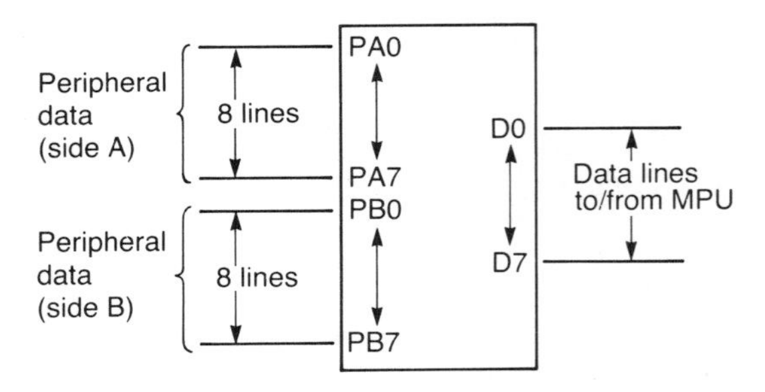

Figure 6.1

port itself. Usually, a logic 1 makes that line an *output* or a logic 0 makes it an *input* and it can be seen that 256 permutations are possible between all lines being inputs only or outputs only.

Some applications require this versatility, but the simplest system is to configure one port to be all inputs and the other all outputs.

Another pattern when placed in the control register will deselect the data direction register and instead select the data register. This device acts as a storage location for data passing either way, into or out from the peripheral chip thus acting as an intermediary between the microprocessor and the outside world.

Also, a range of different patterns in the control register will enable or disable the interrupt system of the device.

Once a port is configured, it is quite often left alone to get on with the job designated to it. It must be remembered that, for each port, the eight data lines are all in parallel and are effectively an extension of the microprocessor data bus.

UNIVERSAL ASYNCHRONOUS RECEIVER/TRANSMITTER (UART)

For some types of peripheral devices, particularly serially driven ones such as a visual display unit

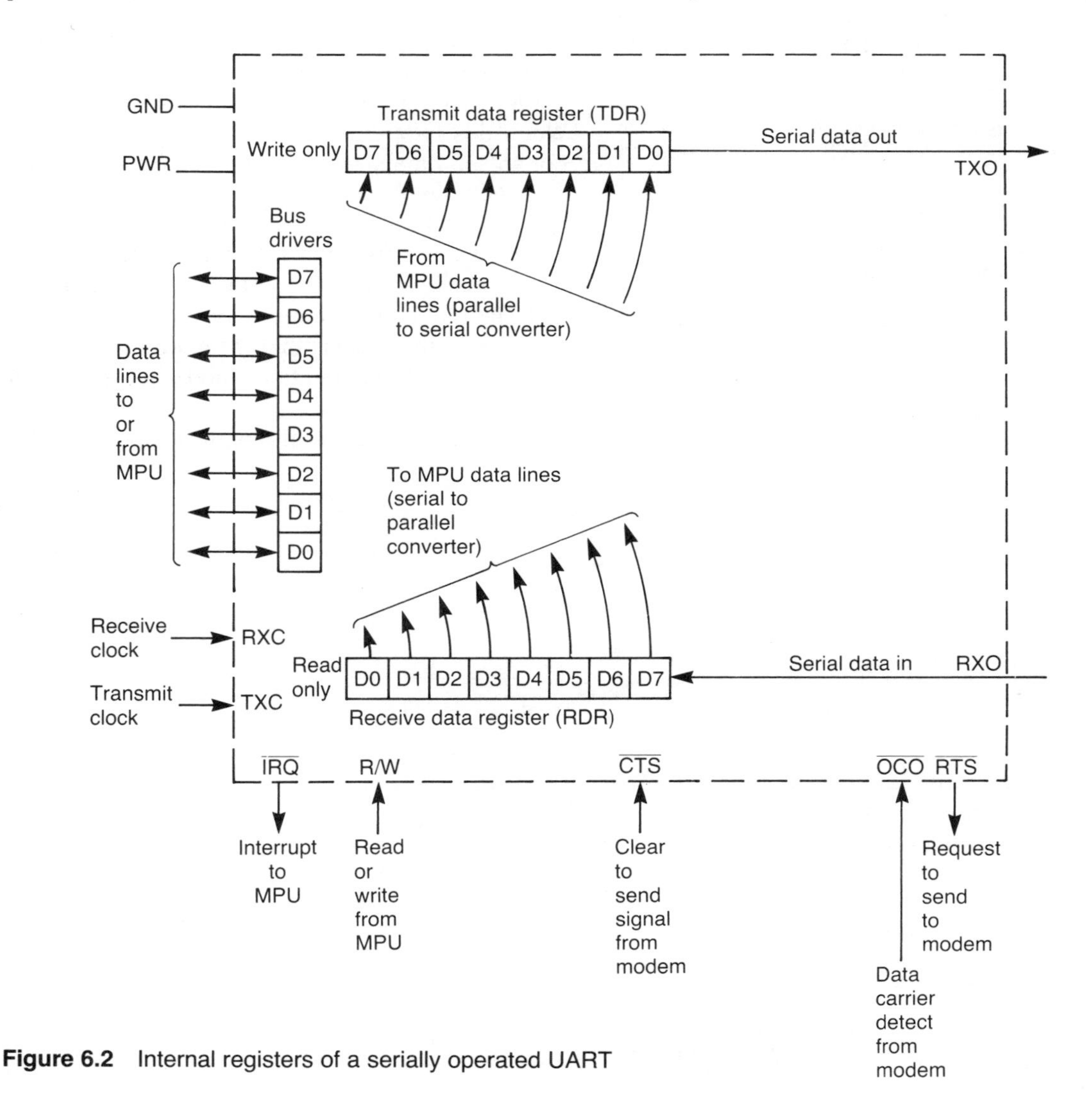

Figure 6.2 Internal registers of a serially operated UART

(VDU), teletype, magnetic tape unit and disk drives, there has to be a form of conversion of the data from *parallel* mode, as it would be on the extension of the 8-bit data bus, into a *serial* form to be sent to and received from the various types of peripheral device previously mentioned. The reason for this conversion lies in the nature of operation of each of the peripheral devices mentioned above. They are all *serially* operated devices and this feature will be examined later.

The block diagram of the UART is shown in Figure 6.2.

It can be seen in Figure 6.2 that a conversion process takes place in the UART where data in parallel form is turned into a serial form. This is carried out by means of special storage circuits called *shift registers*. When a data byte is sent out, this is what happens:

data comes into the UART as eight bits in parallel
data byte is stored in the *data buffer*
this data byte is copied into the *output shift register* as shown in Figure 6.3 below.

Then a sequence of eight timing pulses will cause the data pattern to shift out of the shift register on the serial data output line and leave the system to go to a particular serially operated peripheral device.

Suppose the date byte is 1 0 0 1 1 1 0 0.

Initial pattern	1	0	0	1	1	1	0	0	
After 1st clock pulse	X	1	0	0	1	1	1	0	→0
After 2nd clock pulse	X	X	1	0	0	1	1	1	→0
After 3rd clock pulse	X	X	X	1	0	0	1	1	→1
After 8th clock pulse	X	X	X	X	X	X	X	X	→1

Figure 6.4 Operation of a serial output shift register

Let us look and see how this happens in Figure 6.4 above.

In the diagram X implies that the value of that particular bit after the shift has happened does not matter. It could be either a logic 0 or a logic 1. When this happens it is often referred to in textbooks and data sheets as a *don't care* condition.

Sometimes it is important that the data is not lost, particularly if the peripheral device receiving it finds it is corrupted. It will then ask the UART for a re-transmission of the data byte. If the data is lost then the error cannot be corrected. What happens in this case is that the shift register has a 'wrap around' connection mode as shown in Figure 6.5.

As Figure 6.6 shows, the pattern can be shifted

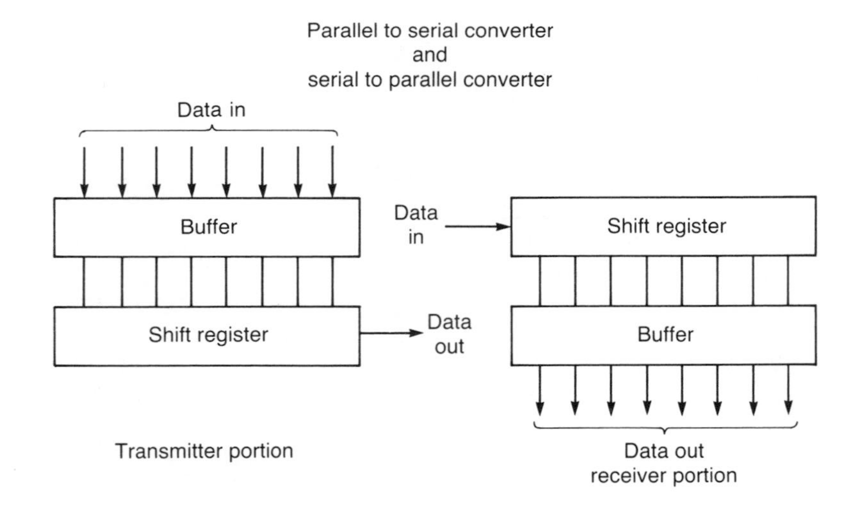

Figure 6.3 Parallel-to-serial and serial-to-parallel converter

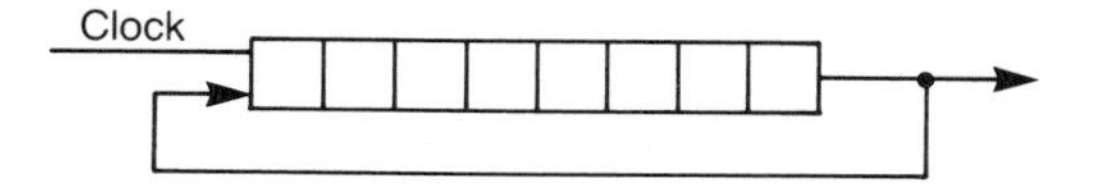

Figure 6.5 'Wraparound' shift register

out from one end and immediately moved back in at the other.

In this way, no data is lost and once the peripheral device is happy that the received data is correct, then the next data byte from the microprocessor can be processed.

Initial pattern	1	0	0	1	1	1	0	0	
After 1st clock pulse	0	1	0	0	1	1	1	0	→0
After 2nd clock pulse	0	0	1	0	0	1	1	1	→0
After 3rd clock pulse	1	0	0	1	0	0	1	1	→1
After 4th clock pulse	1	1	0	0	1	0	0	1	→1
After 5th clock pulse	1	1	1	0	0	1	0	0	→1
After 6th clock pulse	0	1	1	1	0	0	1	0	→0
After 7th clock pulse	0	0	1	1	1	0	0	1	→0
After 8th clock pulse	1	0	0	1	1	1	0	0	→1

Figure 6.6 Effect of 'wraparound' on data pattern stored in shift register

This is obviously slowing down the rate of transfer of data from out of the system and is one of the penalties of using serially operated devices. Fortunately, the problem is not too severe as most peripheral equipment, such as VDUs and teletypes, is serially operated and slow in comparison with microprocessor speed of operation.

CHARACTER REPRESENTATION AND TRANSMISSION METHODS

The microprocessor will handle any data that is presented to it in byte-sized pieces. It has no 'interest' in the data itself and will carry out whatever instructions it is programmed to do. The human user, however, has a great deal of interest in the data itself and it should be in a form which can quite easily be understood.

Basically, when a microcomputer is used in an application such as word processing, spreadsheets, databases or engineering applications such as circuit design, systems design, etc. the human user wants to see *characters* – the letters A to Z in capitals as well as lower case (a to z) and *numbers* – the symbols 0–9.

These characters and numbers are lumped together and called *alphanumerics*.

What the microcomputer user has to do, then, is to use some form of *code* that can represent the alphanumerics in binary so that the microcomputer can accept the binary pattern and handle it in the way the user wants.

There are a number of ways this can be done but one important method you should know about, because it is used in many places in the world and by all the major microcomputer manufacturers, is *ASCII* (pronounced *ass-key!)* – *American Standard Code for Information Interchange*.

ASCII

ASCII can be very conveniently represented by seven binary digits (bits) and is therefore very useful for microprocessor systems using eight bits. The fact that there is one spare bit is a bonus and is used for something else to be explained a little later on.

If you look at a typewriter keyboard, which is often called a 'qwerty' keyboard because of the way the keys are labelled along the top alphabetic character row, you will see many other symbols, for numbers, punctuation marks, brackets, shift and return keys, etc. It has been found that these characters for text and control of text can be represented by 128 ASCII codes.

You will hopefully remember from an earlier chapter that $128 = 2^7$ and so we would use seven bits to represent these 128 codes.

Note A table of ASCII codes is given in Appendix B.

So we now have a method of representing our alphanumeric character set as ASCII binary codes which the microprocessor system would be able to recognise as perfectly normal data patterns.

However, ASCII uses only seven bits. What do we do about the spare or eighth bit? We have two choices: the first is to do nothing and leave it as a zero, i.e. ASCII codes from 00 to 7F *or* we use it in some other way. In practice, the spare bit is often used as a *parity* bit which will be explained a little later on. Just accept for the moment that all eight bits are presented to the microprocessor for it to process in the normal way.

Another problem that has to be sorted out is this; having turned our alphanumeric symbols into ASCII binary codes, what method are we going to use to send them from a serially operated peripheral device such as a teletype or VDU? How will the microprocessor, which is a parallel, clock-driven digital system know when information is being sent?

The VDU and teletype are examples of peripheral equipment most commonly used by human operators. There are many other types of peripherals and it is probably true to say that the VDU has taken over from the teletype for a number of good, practical reasons.

As this equipment is designed for human user convenience, a number of important features are necessary:

(a) the slow speed of human operation must be allowed for
(b) the random nature of human operation must be handled
(c) the capability for two-way, or *duplex*, communication must be provided
(d) the equipment should be able to display the messages being put into the system from the qwerty keyboard (input) and also display the results of any processing that has taken place (output).

INTERRUPTS

All microprocessors have an interrupt system which will allow an outside user to 'attract the attention' of the microprocessor or microcomputer system and then get it to carry out their instructions. The other method of interaction is called *polling* where the microprocessor spends all its spare time asking each of its peripheral interface chips in turn the question 'Does anybody out there in the real world want me to do anything?' Only when a peripheral requests service does the microprocessor stop asking that question and then carry out the instructions required by that particular peripheral device.

With interrupts, a series of priorities is set up at the very start. Some operations are quite important and have to be handled very quickly. For example, the microprocessor could have the contents of the registers corrupted should the power supply fail. It would wish to carry out an action to preserve its register contents, and would give this action the highest priority. Other questions such as 'Which VDU needs work doing?' may be less important.

Rather than interrogate each peripheral device in turn, as in the polling situation, the microprocessor carries on with its normal business, at its normal operating speed, until an outside user at a VDU or teletype hits a key. This starts a sequence of events during which an *interrupt* signal is generated. This is a hardware generated signal and is sent back, via the interface chip itself to the microprocessor. In the flags register is a particular bit for interrupt and if not already set, this will become set to a logic 1. Once the microprocessor has finished the current instruction it is on, it examines the flags to see if anything important has happened since it last looked at them. Now, it 'sees' the interrupt set at a logic 1 and then knows that someone, outside the system, wants it to do something.

It then acknowledges the interrupt, finding out which outside device wants it to do something, such as accept ASCII coded characters into its own system, carries out the necessary tasks and when finished resets the interrupt bit in the flags register to zero. It then continues with the original task.

It is by use of interrupts that peripheral devices can attract the attention of the microprocessor. This initiates a series of instructions which will allow it to know where it left off from its current task when it returns from carrying out the instructions of the peripheral device.

THE RS232 SERIAL INTERFACE

This interface protocol was introduced by the AEIA – the American Electronic Industries Association – as a standard for communication using *modems* between computer systems. A modem (*mo*dulator/*dem*odulator) converts the binary data into audio tones which can then be sent over normal telephone lines.

The RS232 system has come to be used in many guises and is no longer a true standard because the original interface had spare capacity in its hardware that allowed other manufacturers to use the spare capacity in non-standard ways.

With the RS232 Serial Interface, it is usual to have seven bits per character, which is exactly the same as ASCII codes allow. The transmission rate of the bits is in *bits per second* and the unit is the *baud* after the French telegraph engineer Jean Maurice Émile Baudot (1845–1903). A number of transmission speeds are possible but the most common ones used are 75, 110, 300, 1200, 4800 and 9600 bauds.

To allow for the problem of the random nature of human operation, the process is asynchronous, or in other words, *not* controlled directly by the microprocessor clock frequency. This means that each data byte has to be preceded by a *start bit* and then ended by a *stop bit* which may be one or two bits in length:

Start bit – logic 1 – one bit in time duration
Stop bit – logic 0 – one or two bits in time duration
Parity bit – logic 1 – one bit in time duration

PARITY

Parity is a simple but partially effective way of checking a received character for errors. The eighth or 'spare' bit mentioned earlier is used as a 'parity' bit. Depending on whether the chosen parity system is odd or even, the number (quantity) of logic 1's received, ignoring the start and stop bits but including the parity bit, should be respectively an odd or even number.

Odd Parity If the number of logic 1's in the seven bit ASCII data is even in number, the parity bit is set at one to make the total parity of the 8-bit byte odd. If the number of logic 1's is already odd then the parity is set at 0.

Even Parity This is exactly the reverse of above. The total parity of the 8-bit data byte has to be even and the parity bit is set accordingly.

TRANSMITTED DATA

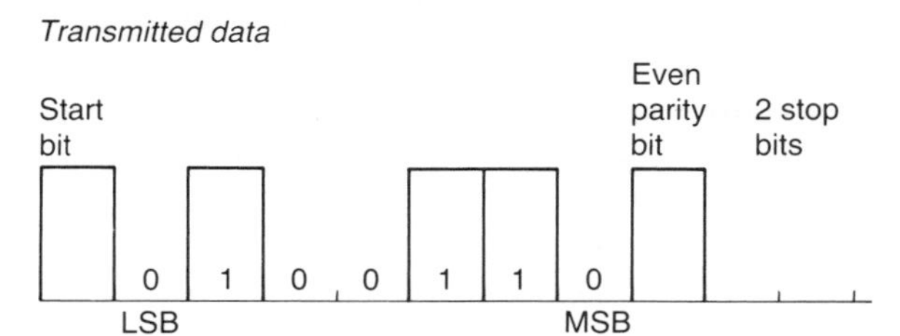

Figure 6.7 Character transmitted by the RS232 system

Figure 6.7 shows a typical RS232 signal of 11 bits in length.

They are:
One start bit
Seven data bits; note the least significant bit (LSB) is sent first
One parity bit
Two stop bits.

Notice from Figure 6.7 the parity system is *even*.

These are all sent at whatever baud rate the system has selected. It can be seen, though, that of the 11 bits sent, only seven have any real information, which slows the system down at the advantage of reducing errors.

HANDSHAKING

Another duty of any communication protocol is ensuring that the message is sent only when the receiver is ready. Figure 6.8 shows how the RS232 uses two signal lines, RTS and CTS, to make sure

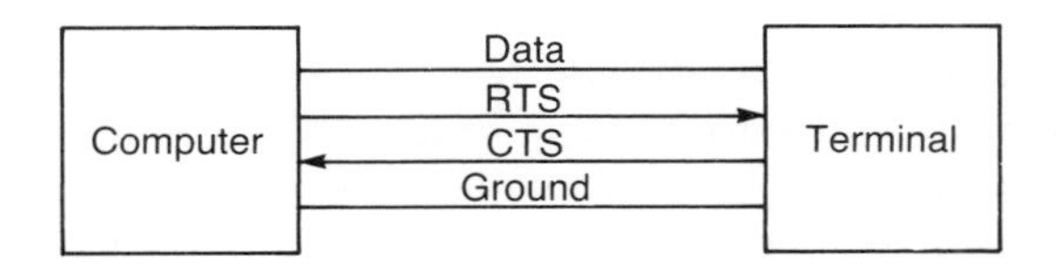

Figure 6.8 Simplified diagram showing 'handshake' signals for the RS232 system

that the communication process is controlled by the slower of the devices, invariably the VDU or teletype. This process is called *handshaking*.

When the computer has data ready it sends the RTS, or *'ready to send'*, signal to the terminal. Only when the terminal is ready to receive it will it send back the CTS, or *'clear to send'*, signal. Only on receipt of the CTS signal will the computer then send one character. In this way the slower device controls the speed of transmission so that no characters get lost.

THE TELETYPE (TTY)

This was an early computer peripheral device and is sometimes called a teleprinter. It is used by many organisations worldwide. It was adopted for early work where transmission speeds were slow but is now superceded by the VDU which can operate much faster because it does not rely on electromechanical devices. A diagram of the teletype is shown in Figure 6.9.

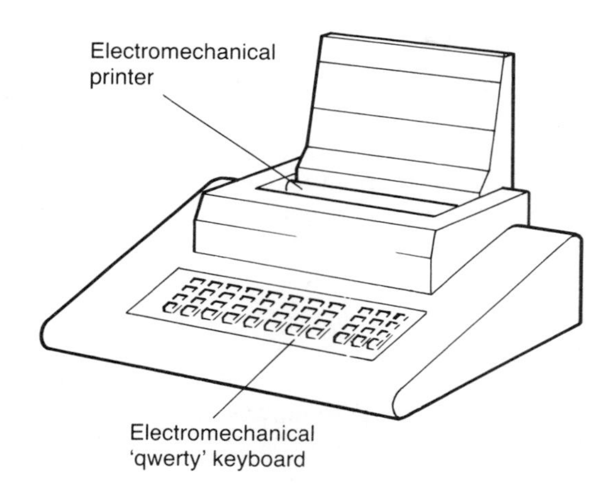

Figure 6.9 Schematic diagram of a teletype

It uses a qwerty style traditional keyboard to allow the operator to input programs and data and produces under local/remote control either hard copy of the operator input messages or the results of output from the computer.

Each character is printed by means of a mechanical typeface, like some typewriters. It is slow in operation and mechanical maintenance of such devices can be a problem.

It is serially operated, asynchronous in nature and so uses the start/stop control elements described earlier.

The keyboard produces the coded characters which are sent down the line to the computer and, in the early days, many different types of code were used.

THE VISUAL DISPLAY UNIT (VDU)

This is the electronic version of the teletype. It uses a qwerty keyboard for input and produces its output on a cathode-ray screen, just like a television. Because it is electronic it is much faster and more reliable than the teletype.

In the keyboard is a ROM which contains the ASCII codes for each of the keys on the keyboard. When an operator presses a key, a location in this ROM is accessed and the binary code is put directly on to the data bus.

The screen is organised in either *graphics* mode or *text* mode. In graphics the resolution is governed by the number of lines of the system, usually 625 lines and resolution comes down to a small area called a *pixel*. A typical medium resolution graphics screen would be arranged to have 1280 pixels horizontally and 1024 vertically. Addressing each of these would take a great deal of memory space which is why graphics processors and screens cost a lot!

To reduce the cost, the resolution could be degraded, and reduced to 640 by 256 pixels where the pixels are now larger areas of the screen. This would take up less memory space.

But *why* should memory space by important? Why does it become necessary to use memory? The reason is that all of the information to be displayed on the screen has to be stored in memory before it is fed out to the screen line by line. The screen is a serial device, with each frame being held for about 20 ms before being updated, or refreshed. So, the more points that are needed, the better the resolution or sharpness of the display, but the greater the number of memory locations that are needed to organise this. All manufacturers produce a cathode ray tube (CRT) controller for use on their systems which carries out all the necessary interfacing to keep the display organised properly.

In the text mode, the same principles apply except that, as each alphanumeric character can be stored in a ROM, the address of each place on the screen where the character is to be placed takes up less space in the CRT RAM area.

In this way, the screen can be organised into the following ways:

80 columns × 32 rows
40 columns × 32 rows
80 columns × 25 rows
40 columns × 25 rows

Other variations are possible, of course, but the above are quite common.

The VDU has taken over from the teletype as a versatile peripheral device to such an extent that all microcomputers sold as home computers or personal computers come equipped with full encoded keyboard and VDU screen as described above. New types of display, such as liquid crystal and plasma displays have reduced both the volume and power supplies needed to run a system. The new generation of lap-top computers use these technologies and can provide many hours of operation from battery supplies.

BACK-UP OR SECONDARY STORAGE MEDIA

Most 8-bit microprocessor systems have a total memory map space of 64 K or 2^{16} bytes of memory which has to do everything, including the monitor program in ROM and VDU screen in RAM. What space is left is available both to the system and user to run programs or applications.

Because RAM memory is volatile, which means all data gets lost when the power is removed or broken for even a few microseconds, and the RAM space is often too small, most serious users have their programs and data on secondary storage devices such as cassettes or floppy disks.

Under operating systems control programs can be *loaded* into RAM from the cassette or disk and then run normally. Once the program has run the new results can be *stored* back on to the secondary storage media for safety. If the system 'crashes', or fails for any reason at least all is not lost and the last version of the program is safely stored.

CASSETTE STORAGE

As most of you are well aware cassettes are very versatile devices for storing music, taping music from the radio, playing commercial tapes of various kinds, etc. Most tapes are of 30, 60 or 90 minutes duration, divided between the two sides of the tape.

They are also useful for the storage of computer data and programs but certain precautions need to be taken and modifications made for them to work properly.

Firstly, tape is very definitely a serial medium. It starts at one end and plays until it reaches the other end, when it stops (or reverses). If a computer is to use this medium it can be painfully slow to find and load a program.

Secondly, computer data is in binary form, 0's and 1's, which have to be converted so that they can be recorded on the cassette tape as it moves across the heads.

Thirdly, as it is a serial process, a peripheral interface chip has to be used to process the serial data, store it and then pass it to the microprocessor in parallel form for it to be processed.

The serial nature of a tape storage system causes us to look at ways of improving and speeding up the process. The first thing is to invent a filing system, under which our programs and data can be stored properly.

In a microprocessor, most instructions and data are in binary digits or *bits* which is understandable to the microprocessor but *not* necessarily to us! We like to call our files by *name*:

e.g. FRED 1, FRED 2
BILL 1, BILL 2, BILL 3, . . .
ANNE 1, 2, 3
SUE 1, 2, 3, 4, etc.

or PROG1
PROG2
LOOP3
ROUTINE4 etc.

We use the *name* to identify the start of that particular program file which is going to be stored somewhere on the cassette tape. When we give the operating system a LOAD command through the keyboard, the system will ask for the NAME to be found and loaded.

For example, *LOAD PROG1*; the system responds

'SEARCHING FOR PROG1'
'PROG1 FOUND'
'PROG1 LOADED'

or some other variation to inform you that the file has been found.

The problem is of course, when the Play button on the cassette recorder is operated, it can take ages before the file is found and this is very annoying! A way of overcoming this is to have control of the Fast Forward motor and run the tape until the file name is found by the system. Once found the motor speed changes to normal speed and the program is loaded.

It can still take two or three minutes to run through a normal tape, though, and one remedy was to produce tapes of five minutes duration for computer users as most programs will be loaded in seconds, rather than minutes, unless they are sophisticated and lengthy programs.

The logic 0's and 1's are converted into audio tones of 1200 Hz and 2400 Hz, well within the audio range of the cassette system. The microcomputer system converts the serial data stream of 0's and 1's into these tones which are then stored onto the cassette tape in exactly the same way as music would be from a radio. The additional circuitry used for this purpose usually has one of two names:

CUTS – Computer Users Tape Storage, or
KANSAS CITY – after the town where an interface was specified for this purpose.

Most microprocessor systems managers will use one or the other of these interface standards.

The microprocessor handles the problem of serial-to-parallel conversion in much the same way it handles other serial devices. The UART peripheral chip described on page 51 has been designed for parallel-to-serial conversion and to adopt whatever transmission speed and protocol desired. Once this is set up in a particular system it is usually left well alone!

DISK STORAGE

IBM produced an early floppy disk drive format, derived from one of their larger fixed disk systems used in mainframe computers. The technology has developed to such an extent that, just as the VDU has taken over from the TTY, so the floppy disk has rendered cassette systems obsolescent.

The disk is run at a speed of about 360 r.p.m. and the surface is originally unmagnetised or organised in any way. The first operation is to *format* the disk which means to lay down or arrange the magnetic surface into a series of concentric circles called *tracks*. The 8-inch disks have 77 tracks, 75 available to the user. The $5\frac{1}{4}$-inch disks which were subsequently developed have either 40 or 80 tracks and could be either single-sided or double-sided, which increased their capacity enormously.

Additionally, from single-sided, single-density 8-inch floppy disks which could store 256 kbytes new $3\frac{1}{2}$-inch double-sided, high-density mini floppy disks are available, which can store up to 2 Mbytes!

These 8-inch disks are made from thin sheeting such as MYLAR, coated with an iron oxide which is easily magnetised. The disk is contained in a cardboard envelope with a soft tissue lining to protect the disk surface and keep it clean. It looks a bit like a large 45 r.p.m. record, except that it stays in the cover! Single-sided or double-sided disks are available in single-density or double-density formats. The meaning of single- or double-sided is obvious, but single or double density needs explaining. Single-density recording is the way in which data is stored on the magnetic surface of the disk. Figure 6.10 shows how each data cell is set up by means of clock pulses. A clock pulse is timed to occur at the start of each data bit in the character. Depending on whether it is a logic 1 or logic 0 determines whether there are two pulses or one stored in each cell.

So, each data cell has the clock pulse present and this is a particular feature of the single-density method of recording data.

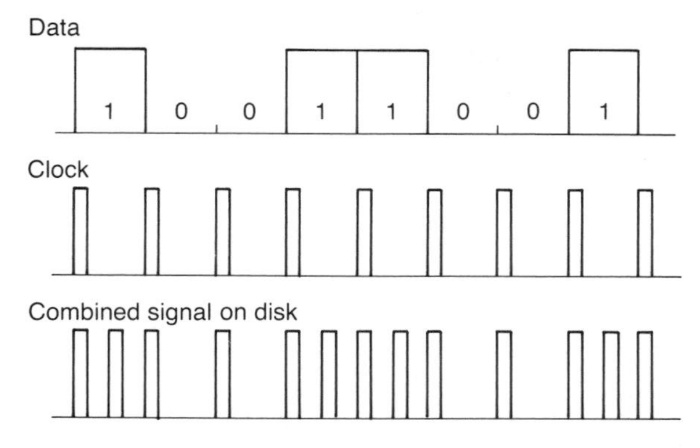

Figure 6.10 Diagram showing how data bits and clock pulses are combined with data cells on the disk surface in a single density recording system

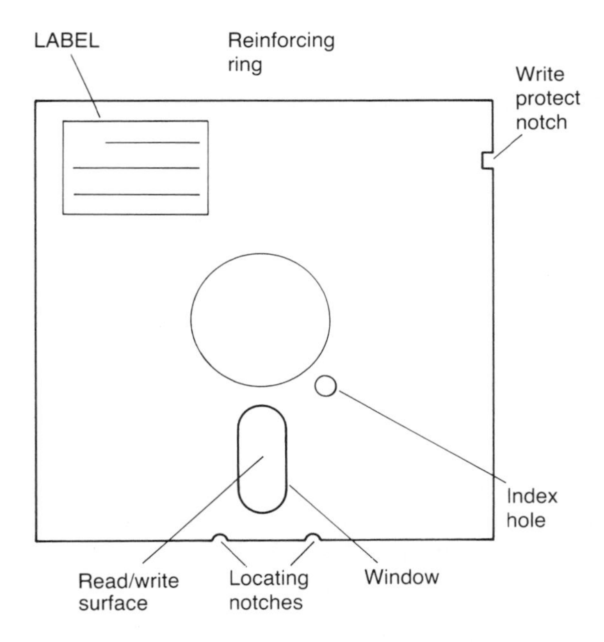

Figure 6.11 External features of a 5¼ inch floppy disk

Double-density recording is a way of increasing the data that can be stored on the disk, allowing for a possible doubling of the data rate.

Figure 6.11 shows the external features of a $5\frac{1}{4}$-inch disk.

The surface of the disk itself is covered with magnetic material which, when formatted, means that it is organised into 80 concentric tracks numbered from 0, the outermost, to 79, the innermost. This is shown in Figure 6.12.

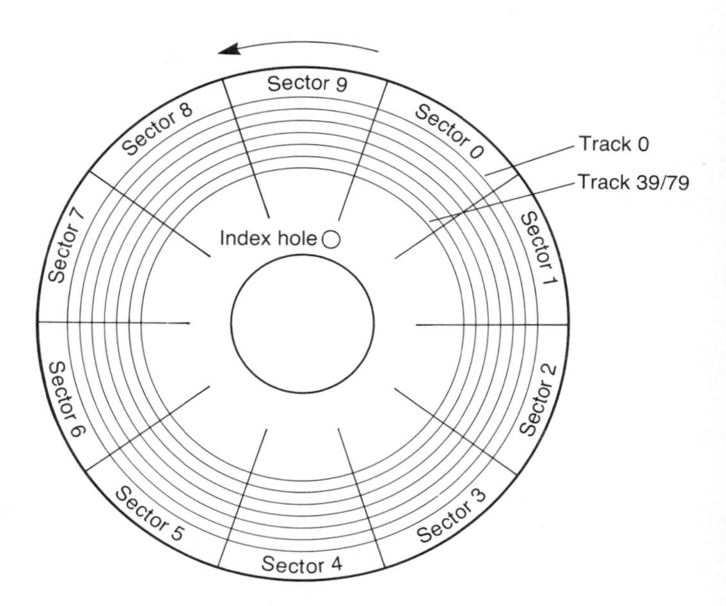

Figure 6.12 Track and sector layout

To increase economy of use, each track is further divided into *sectors*. Typical numbers of sectors are 8, 15 or 26, but other possibilities are available. Most commonly used is '*soft sectoring*', where the disk operating system organises the number of sectors per track, as opposed to *hard sectoring*. Soft sectoring is so called because it is organised by the software, or computer programs written for the disk operating system. Hard sectoring is carried out by the disk manufacturer drilling holes at fixed intervals in the disk surface.

Because the tracks are concentric, the read/write heads have to know where Track 0 is and where the start of each track is. The diagram in Figure 6.13 shows the basic arrangement of a disk drive and its interface with the microprocessor system.

In Figure 6.13 a 'Track 0' detector is in place to ensure that the rest of the system is aligned properly from Track 0.

The *index hole* is a mechanical indication of the start of each track.

Once the system is set up properly disk read and disk write operations *should* be simple, fast and effective!

Other technologies, such as the Winchester disk drive, have been developed which allow for the storage of data and programs and have capacities

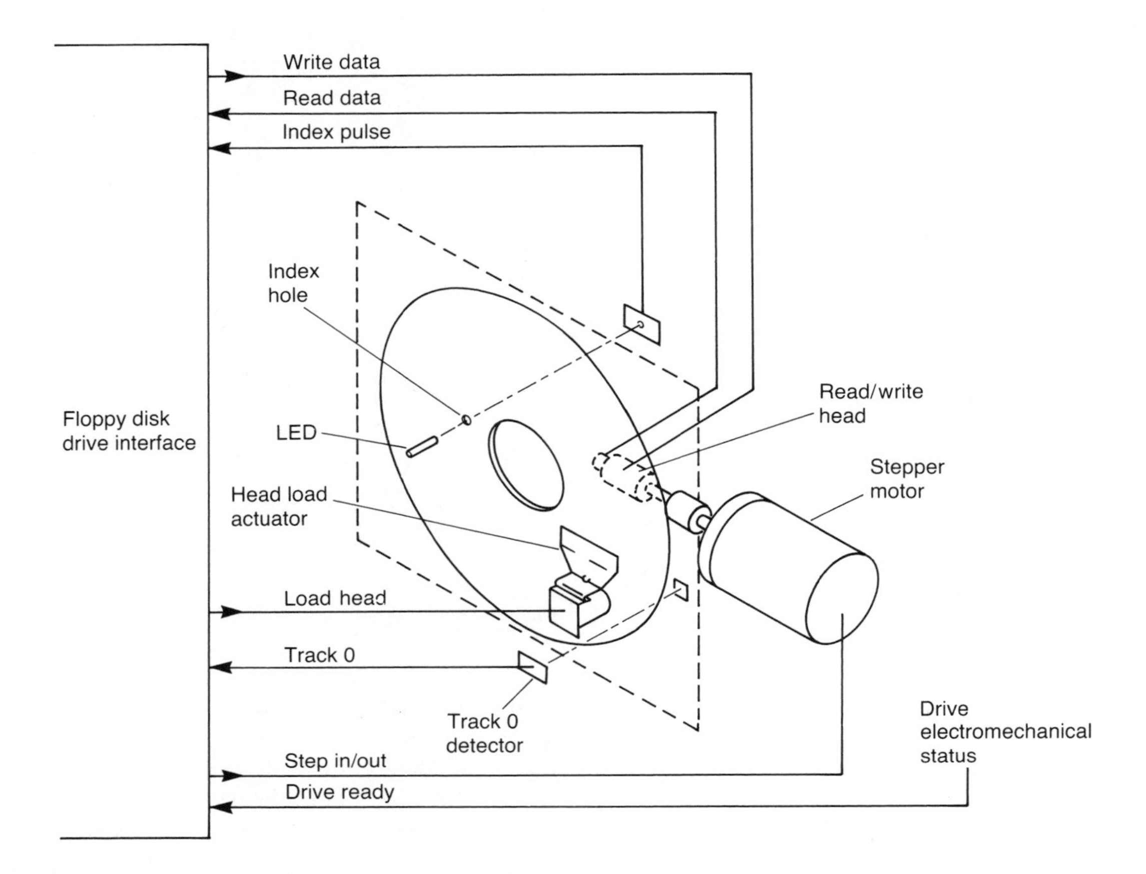

Figure 6.13 Main features of a floppy disk drive

typically of 20 MBytes, 40 MBytes, etc. up to 120 MBytes! These will not be described here but the point is that more and more storage capacity is available in increasingly smaller volumes.

Summary

This chapter has seen how the microprocessor system is changed into a microcomputer system by the addition of external pieces of hardware. These are intended to make operation of the system easier for the human user, particularly users who have little interest in programming as such, but want to use the system to run a particular application such as a wordprocessor or a game.

These additional pieces of equipment or *peripherals* as they are commonly called have many functions and uses but the four examples that have been considered are:

the *teletype* and *cassette*
and *VDU* and *disk drive*

with the latter combination being the most commonly used one nowadays.

These peripherals have to be connected to the microprocessor system itself and the interfacing of them to the data and address buses needs special interface chips and programming to ensure they do what the user wants. It is in this interfacing, most of which has already been done by others for us to use, that most of the interesting work of a microprocessor system lies.

Suggested Assignment Work

1 Examine manufacturers' data sheets and find out the chip details for interfacing Motorola and Intel 8-bit microprocessors to *VDU*, *disk*, *teletype*, *cassette*.

2 From these data sheets try to discover the methods used by the manufacturers to interface the slowness of the peripheral to the high speed of the microprocessor.

3 Examine the interrupt strategies of the Motorola 6802 and Intel 8085 microprocessors and compare them.

4 Use a microcomputer system which has VDU and disk drive. Examine the disks and see how they are loaded into the system.

5 Examine a *disk filing system* to see how it is organised.

6 Find out the meaning and use of a 'hierarchical' filing system.

SEVEN

Hardware, software and firmware

> **OBJECTIVES**
>
> *At the end of this unit all students should be able to:*
>
> *Explain the meaning of the terms hardware, software and firmware, and determine the components asssociated with software and firmware.*

The previous chapters have presented some of the elements or component parts of a microprocessor based system and have shown how these can be expanded into a microcomputer system by the addition of peripheral devices such as a VDU and a disk drive.

We have seen that a microprocessor system needs external memory chips such as ROM and RAM to make it into an effective system and, briefly, mention has been made of the fact that the operating system, or monitor routine as it is sometimes called, is placed in, or stored in a ROM which is non-volatile memory. (Remember, this means that the contents of the memory are not lost when the power is switched off.)

A user's program is often entered into and stored in RAM so that it can be run, changed, modified and then rerun until the person developing the program is satisfied with the situation. (Remember, RAM is volatile memory and data is lost when power is removed.)

All these processes and the components in the system which support them, or more simply, allow us to use the system, can be described in more than one way.

HARDWARE

This term is a general one, describing all the tangible parts of the system. This includes the box it is contained in, the printed circuit boards, all the various chips and power supplies, together with all the peripherals.

FIRMWARE

When the hardware of a system has been built it is sometimes referred to as an 'empty' system. This means that whilst it is capable of running programs, applications, tasks, etc. it has not been 'organised' to do so.

The manufacturer has provided very limited ways of gaining access to the microprocessor itself, usually by means of the interrupt system and external control signals.

To make sure that the system is more 'user friendly', the manufacturer will have written a simple form of computer operating system, reduced in scale and complexity, to suit a microprocessor. The basic requirement of this simplified operating system, or 'monitor program' as it is sometimes called, is to check continually if a user is operating the key-pad as an input and then to carry out some form of action as a result of that key-pad input.

For the present, just assume that the monitor program is present and running, which will allow user programs and data to be run. These user programs have to be written in a way that a microprocessor will be able to recognise, decode and then to carry out, or execute them.

The instructions that a microprocessor can recognise are usually very limited, such as:

LOAD	OR
STORE	INCREMENT
ADD	DECREMENT
SUBTRACT	COMPLEMENT
AND	

What the manufacturer does is to shorten these instructions, which are all using different numbers or letters, into the things called *mnemonics* (pronounced *nemonix*) which hopefully will convey to the user a simple idea of what the instruction does:

e.g. LOAD

This operation has the effect of placing a binary number into one of the internal registers of the microprocessor, usually the one called the *accumulator*. Each manufacturer uses different styles and mnemonics. Some examples include:

(i) For Motorola:

LDA A Load accumulator A.
LDA B Load accumulator B.

As the Motorola chip has two accumulators, A and B, two separate mnemonics are necessary.

(ii) For Intel:

LDA Load accumulator.
MVI Move a data pattern of 1 byte into a single register.
LXI Load a register pair with two bytes.

STORE

This is the opposite of LOAD. A data pattern, already in a register or the accumulator, has its contents copied into a memory location.

(i) For Motorola:

STA A Stores contents of accumulator A in memory.
STA B Stores contents of accumulator B in memory.

(ii) For Intel:

STA Stores contents of accumulator in memory.
SHLD Stores the contents of the double register HL in two memory locations.

ADD

This is the addition instruction working in binary. It may or may not involve the value of the *carry* flag bit, depending on how the user writes the program.

(i) For Motorola:

ADD A Add the contents of memory into accumulator A.
ADC B Add the contents of memory into accumulator B, but also taking the carry flag into account.

(ii) For Intel:

ADD r Add the contents of register r into the accumulator.
ADC r Add the contents of register r into the accumulator, but also taking the carry flag into account.

SUBTRACT

This is the subtraction instruction, working in binary. This may or may not use the carry flag bit in place of a '*borrow*'.

(i) For Motorola:

SUB A Subtract the contents of memory from accumulator A.
SBC B Subtract the contents of memory from accumulator B, also taking the carry flag bit into account.

(ii) For Intel:

SUB r Subtract the contents of register r from the accumulator.
SBB r Subtract the contents of register r from the accumulator with a 'borrow'.

AND
This is the instruction that carries out the Boolean logic AND function. It works on corresponding bit positions in the accumulator and memory.

(i) For Motorola:
AND A Boolean AND between memory and accumulator A.

(ii) For Intel:
ANA r Boolean AND between a register and accumulator.

OR
This is the instruction that carries out the Boolean logic OR function. It works on corresponding bit positions in memory and the accumulator.

(i) For Motorola:
ORA A Boolean OR between memory and accumulator A.

(ii) For Intel:
ORA r Boolean OR between a register and accumulator.

INCREMENT
This is the instruction that causes the contents of an accumulator, a register or a memory to increase by one.

(i) For Motorola:
INC A Increases contents of accumulator A by 1.
INC Increases contents of memory by 1.

(ii) For Intel:
INR r Increases contents of register by 1.
INX B Increases contents of double register B and C by 1.

COMPLEMENT
This is the process, sometimes called *inversion*, of changing the value of each individual bit position in the accumulator, i.e. a logic 1 becomes a logic 0 and vice versa.

(i) For Motorola:
COM A Complements the contents of accumulator A.

(ii) For Intel:
CMA Complements the contents of the accumulator.

Even these abbreviations are not understood by the microprocessor and, as a necessity, the manufacturers issue tables of instructions or op-codes which are used to convert the mnemonics into binary patterns. Most manufacturers convert them into hexadecimal for convenience. (A comparison of denary, binary and hexadecimal is made in Appendix C.)

As mentioned previously, the manufacturer has to use the instruction set of its own microprocessor in just the same way that any user would. Having written the instruction set it is highly unlikely that it would be modified at a later date.

This 'operating system' or 'monitor program' would allow the user to have access to the microprocessor instruction set as part of a system, which would include input facilities such as a simple key-pad or a qwerty keyboard, and output facilities to a seven segment display or a VDU screen. The capability of storing user programs on disk and loading them into the system memory should also be included.

It is quite common for a manufacturer to provide this operating system or monitor program in ROM which means that it cannot be changed by the user, or corrupted by outside events or lost when power is removed. Under these conditions the operating system software has been developed to such an extent that the manufacturer is content that it will do its job properly for a significant period of time and would like to 'set it in concrete', i.e. ensure that it cannot be corrupted or erased.

This is where the term 'firmware' really derives from: it is operating system software that is to be kept fixed for a longish period of time and which allows all users exactly the same facilities.

Hence firmware is software that is set into ROM. The impression has been given that it is only the manufacturer that uses firmware but this is not true. Any user who has developed software to an extent that they want it stored in ROM, or a variant of ROM, will call it firmware. So, two types of firmware can exist: manufacturer's monitor programs and user's programs.

What types of ROM devices then can be used to do this? Can the user have the same type of ROM as the manufacturer?

To answer this, one or two further terms must be mentioned. They are part of the ROM family but there are significant differences between them.

(a) *ROM – Read Only Memory* – is the family name for non-volatile memory.

(b) *PROM – Programmable Read Only Memory* – is a name given by the semiconductor manufacturers to the devices they produce in their own factories. These devices are produced by using a variety of masks and chemical processes – a process described in Chapter Four. The manufacturers will pay the cost of making these chips this way because they know that thousands of them, all identical to each other, can be produced in one production run. Once the manufacturer has programmed them, however, the user is powerless to change them.

(c) *EPROM – Erasable Programmable Read Only Memory* – is a very versatile device much used by system designers and system users to store *their* own developed programs when they are happy with them. An EPROM can be 'wiped clean' by exposing the semiconductor surface to ultra-violet light through a clear, but sealed aperture, or hole, in the chip itself. An EPROM programming device can then be used to transfer the user's program to the EPROM by using voltage pulses of between 15 V and 25 V to set or reset the individual memory devices in the EPROM structure. Once this is done properly the contents will remain fixed until the user wants to change them, once again, by passing it into the ultra-violet erasing box.

In this way the user has a versatile ROM device in which programs can be stored safely but can be changed at some future time if necessary.

This gives another view of the term 'firmware', as software which is not set for all time, but can be modified in the future. By using PROMS, however, the manufacturer does not have that capability; once 'burned-in', any changes result in those chips being wasted and thrown away.

SOFTWARE

Quite simply, this is the collective name given to all sorts of other programs, some of which will be explained further. The one common element is that *software* is the name given to programs which are loaded into RAM. As previously explained, any power supply loss, however temporary, is liable to corrupt the program, which will then have to be re-entered into the RAM again to get it to work properly.

These programs are commonly of two types.

The first is the very straightforward programs that users write to make the system carry out some function for them. They will use exactly the same instruction set as the manufacturer and will use mnemonics and hexadecimal op-codes in the same way as before, but these will be entered into RAM. Usually these programs are short and simple to enter so that, when the power is removed, they would not take too long to enter by hand once again. The beauty of such programs is that they can be changed very easily indeed – a great advantage during program development.

The second is the use of applications software such as wordprocessor, spreadsheets, etc. which come to the user on a cassette or, more likely now, a floppy disk which has to be loaded into a disk drive. The monitor program can then be instructed to copy the contents of the program on the floppy disk into the RAM. The program then can be run from the RAM and any results kept in RAM.

Quite often, users will have written their own software which, if lost or corrupted, would take a long time to re-enter. In this case, they can *save* their software in RAM on a floppy disk. If this is done at regular intervals it can be kept updated so that, if a power supply failure occurs, then all is not lost and the program can be *loaded* from the floppy disk to allow the development of the program to continue. This, in fact, is not only a sensible way to proceed but has become an accepted part of the way a system is used.

MICROPROCESSOR INSTRUCTION SETS

The microprocessor itself, as a programmable integrated circuit, will respond to a series of programmed instructions if they are presented to it in a logical fashion and if it can understand those instructions. If this is the case then those instructions will be executed, or carried out, and as a result some task will be completed, some processing work done or some application carried out.

The microprocessor will obey a series of instructions if presented to it in the correct way. Earlier in this chapter some mnemonics were introduced which in a very simple way have some meaning for human users but are not directly intelligible to the microprocessor itself.

An *instruction set* is a list of, or table of, conversions between the mnemonics we understand and *hex codes* which the microprocessor can understand. Remember that hex (or hexadecimal in full) is a shorthand way of representing four binary digits or bits. This means that two hex digits are needed to represent an 8-bit instruction. Because there are eight bits in an instruction it follows from earlier work that there can be a maximum of 256 separate instructions.

Before some of these instructions are examined in any detail one or two concepts concerning programming (or using the instruction set) have to be explained. The ideas to be introduced are used in many different computer systems, both small and large and it is important that these be worked at until they are properly understood.

To lead into this, consider the very simple instruction below which we hope the microprocessor would get right!

Load the accumulator with a number (say, 2)
Add to the accumulator another number (say, 3)
Store the result somewhere else (5)

This is a very simple program!

However, it raises some questions, such as:

How does the microprocessor know where to find the first number from?
How does the microprocessor know that it has to be put into the accumulator?
How does the microprocessor know where to store or save the result?

To start to answer these questions we need to understand a number of topics such as:

memory map of the microprocessor
addressing modes of the microprocessor
instruction types of the microprocessor.

MEMORY MAP

Both the Intel and Motorola 8-bit microprocessors use a 16-bit address bus and we know now that gives us 2^{16} separate addresses, i.e. 64 kbytes of memory, or 65 536 bytes ($64 \times 1024 = 65\,536$).

The sixteen address bits can be represented in hex by four digits ranging from

0000 (= 0) to
FFFF (= 65 536).

It has been suggested that our microprocessor has access to a certain number of separate memory locations and each location is a certain size. In fact *all* locations are byte sized and there are 65 536 of them. The microprocessor knows where each address is because, just like house numbers along a very long road, each memory location in the microprocessor system has its very own 16-bit address.

This is represented diagrammatically by a *memory map* which looks like a large number of piegeon holes in a library system, see Figure 7.1. In each box, pigeonhole or slot will be one byte, or eight bits of information.

It is obviously not practical to draw *all* 64 K of memory locations so we only draw the ones we need for a given job or application. It can often be simplified further by referring to the addresses in groups such as RAM addresses, ROM addresses, EPROM addresses and so on. The diagram could be redrawn as shown in Figure 7.2.

It is no accident that in both Intel and Motorola systems ROM is placed at the high numbers and RAM at the low numbers of addresses and this is

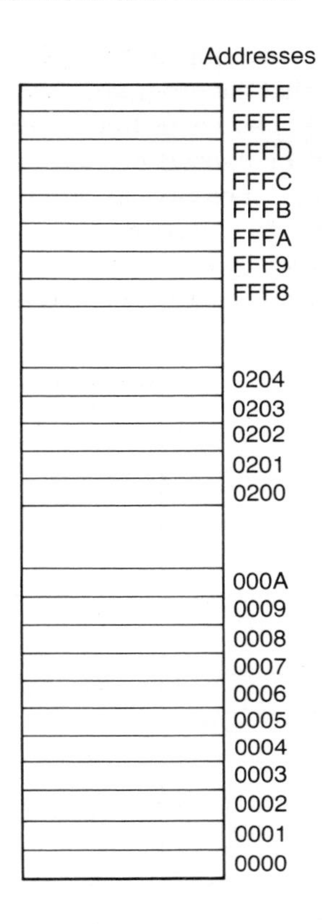

Figure 7.1 Memory map for a microprocessor with a 16-bit address bus. The addresses, when expressed in hexadecimal (hex) abbreviate to four hex characters

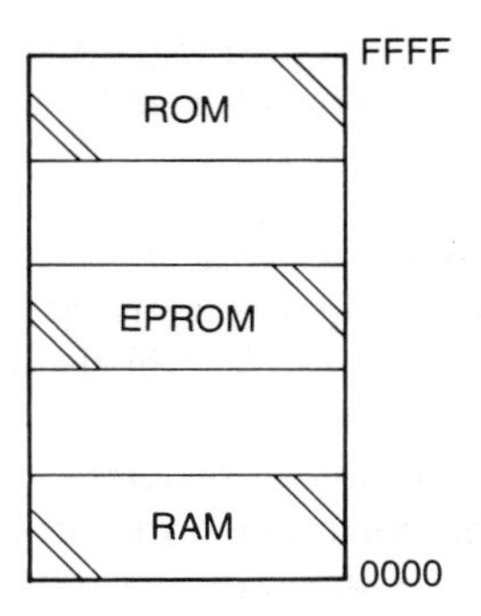

Figure 7.2 Simplified memory map for a typical mircroprocessor system, showing where ROM, EPROM and RAM could usually be expected to be placed

quite often used by other microprocessor manufacturers. Remember the *monitor* program is usually in ROM and is situated at the top of the memory map for reasons that will not be discussed here.

User firmware stored, or resident, in EPROM can be placed anywhere else in the memory map but it is quite common to place users' programs and data in RAM at the low address numbers end of the memory map.

Exactly *where* each type of memory chip will be placed and *how much* memory space each one takes up are questions which are best left to the system designer. Just accept the fact that these addresses are fixed and we, when we use the microprocessor system, have to be told where these addresses are, usually by referring to the manufacturer's manuals.

ADDRESSING MODES

Having introduced the concept of the memory map and also having seen that RAM for users' programs and data *usually* occupies the lower part of the total memory space, what now has to be done is to introduce the various ways that programs and data can be separated from each other and stored in the RAM area.

Remember, RAM is often used to develop and try programs but is also used temporarily while applications programs are being run.

The way in which a program developer uses the RAM area of the memory map is affected by the addressing modes of the microprocessor that is chosen to be used. Let this idea be considered by referring again to the simple program mentioned earlier on:

Load accumulator
Add to the accumulator
Store the result

The instructions – Load, Add and Store – when converted into op-codes or hexadecimal instructions to the microprocessor tell it a number of things. An op-code tells the microprocessor *what* has to be done, *where* the information is that is to be processed and *how long* the instruction should take in terms of the number of clock pulses or cycles needed. This is all part of the fetch–execute cycle described in Chapter Five. It also allows the introduction of two other words that you need to know about:

Operation – the same thing as an instruction – something that the microprocessor will do to carry out the instruction.

Operand – the 'thing' that is going to be 'done unto'! It can be data or another address and the difference between the two can often be very worrying and confusing. The microprocessor is not confused because the *addressing mode* that the program selects will fix this for it.

We will not look at these addressing modes in any more detail but hopefully you will have enough information to allow you to write some simple programs of your own.

The Intel 8085 microprocessor uses four different addressing modes, which are:

Direct
Register
Register indirect
Immediate

Motorola, in the MC6802 microprocessor, uses seven different addressing mdoes, which are:

Accumulator
Immediate
Direct
Extended
Indexed
Implied
Relative

Note
Intel 8085 Direct = Motorola 6802 Extended
Intel 8085 Immediate = Motorola 6802 Immediate

Apart from this there is very little direct comparison because the architecture of the two microprocessors is very different. The major similarity lies in the fact that both use an accumulator register to store results of operations and processes and this is the main general purpose register shared both by the microprocessor and the programmer.

When the microprocessor 'fetches' an instruction and then 'executes' it, the microprocessor has to 'pointed' in the right direction to know where to start from, just as many humans do.

Remember that the program counter referred to in an earlier chapter keeps track of the program as it is executed or carried out and 'points to' the address of the memory location of the next instruction to be carried out. This will be explained further by examining the memory map for the RAM area as shown in Figure 7.3.

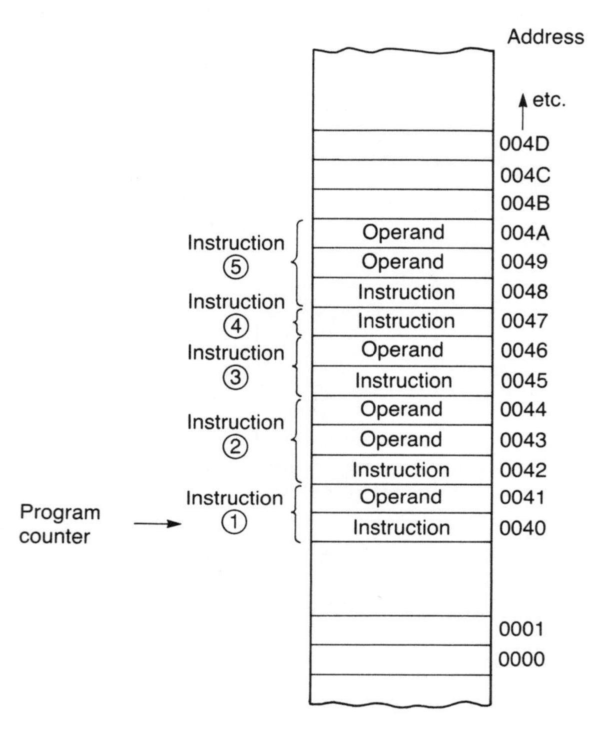

Figure 7.3 Memory map showing INSTRUCTIONS of various lengths and their OPERANDS. Note that each follows on in consecutive addresses from its predecessor

Figure 7.3 shows an area of memory occupied by a RAM of a certain size, ranging from, say, hex addresses 0000 to 007F. We will also assume that a simple program is to start from memory location 0040 and that this simple program consists of five different instructions. Once the program counter is loaded with the start address, 0040, the system will do the rest and the program counter will move to addresses 0042, 0045, 0047 and 0048 in sequence until the program is finished.

How does the microprocessor *know* how to do this? It does so because each instruction, represented by a hex op-code contains all the relevant information about where the operand is, how many RAM memory locations are needed for that instruction and how many clock cycles are needed to carry it out?

You will have noticed also that the memory space used by instructions and their operands can vary from *one* byte of memory (1 location) to *two* or *three*. In fact we often call our instructions one, two or three bytes in length of memory space as a convenient way of classifying them. This means that some instructions need no operand and are self-contained while others have operands that are either one or two bytes in length.

The reason for this is that the operand can be a byte of data or two bytes if it is an address, which is 16 bits in length.

Other possibilities exist, but these are the two main types that operands can be. This needs further explanation and once again, the memory map will be used.

In Figure 7.4 the memory map shows how memory space in RAM is organised to store a program and the data that program needs.

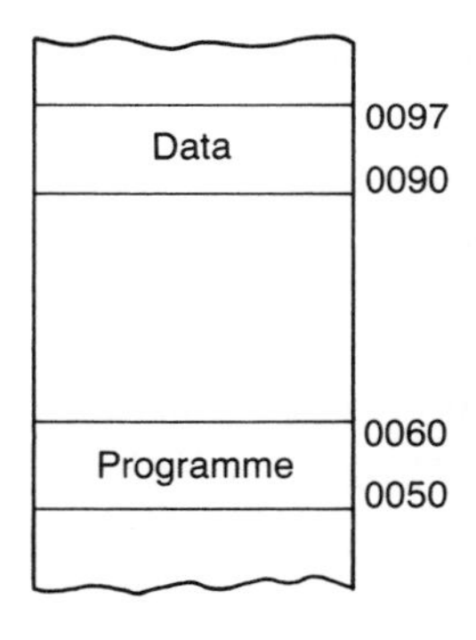

Figure 7.4 Memory map showing that a PROGRAM and its DATA needed are placed in separate areas in RAM

In this case the data is stored in an area some distance away from the program and here the operands for the instructions would be the relevant addresses in the range 0090 onwards, for example.

We can now refer back to our simple program:

(a) Load accumulator with a number (say, 2). Assume that memory location 0090 has the number 2 stored in it.
(b) Add to accumulator another number (say, 3). Assume that memory location 0091 has the number 3 stored in it.
(c) Store the accumulator contents (now 5) into a memory location. Assume this is 0092.

In this case the instructions would have to have as their operands the addresses 0090, 0091, 0092, and the program could be written in addresses 0050 onwards, as shown in Figure 7.5.

Address			
	0050	Load accumulator	
	0051	From address (1)	0090
	0052	From address (2)	
	0053	Add to accumulator	
	0054	From address (1)	0091
	0055	From address (2)	
	0056	Store number in accumulator	
	0057	From address (1)	0092
	0058	From address (2)	

Figure 7.5 Example program which LOADS the accumulator with a number from memory, ADDS a number to it from another memory address and finally STORES the result in a third memory address

In this example the memory map could look like that shown in Figure 7.6.

Location 0092 will have its contents overwritten by this program after it is run and should have the number 05 stored in it.

The numbers and addresses can be entered into the RAM but the instructions need changing into hexadecimal op-codes. In the above example the op-codes for either the Intel or Motorola microprocessor would be as shown in Figure 7.7.

Note the difference in the way the address is specified.

In this example, Intel would be using their Direct addressing mode and Motorola their Extended addressing mode; different names and different op-codes but each effectively doing the same thing. The memory maps for the two processors after the programs have been run would now look like those in Figure 7.8.

Both processors use the same amount of

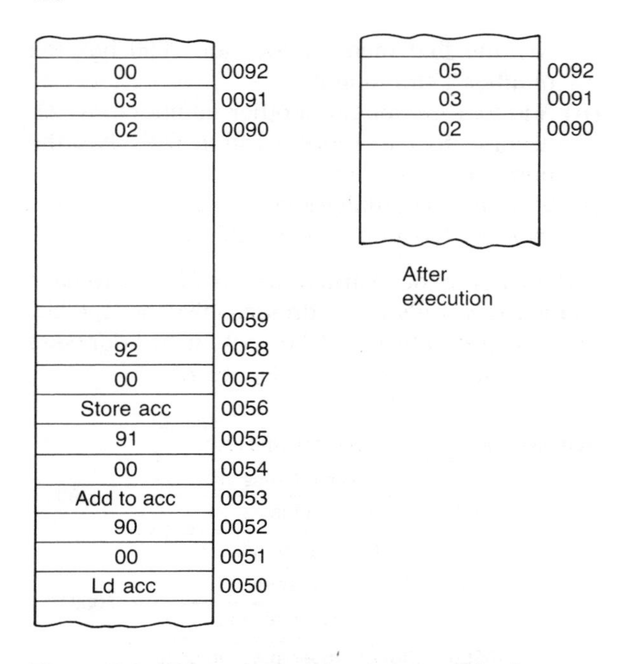

Figure 7.6 Memory map showing possible changes to memory location contents as a result of carrying out the program in Figure 7.5

	Intel	Motorola
Load accumulator from address 0090	3A 90 00	B6 00 90
Add to accumulator from address 0091	86 91 00	BB 00 91
Store accumulator contents into address 0092	32 92 00	B7 00 92

Figure 7.7 Comparison of Intel and Motorola machine code programs to LOAD, ADD and STORE

memory space but using different op-codes produce the desired result.

The method used above of keeping the data in a different area of RAM, well away from the program itself, appears awkward but has the advantage of allowing us to change the data whenever we want to without having to change the program.

Sometimes that is not necessary and by using the Immediate addressing mode of either processor the data can form part of the program itself.

For example,

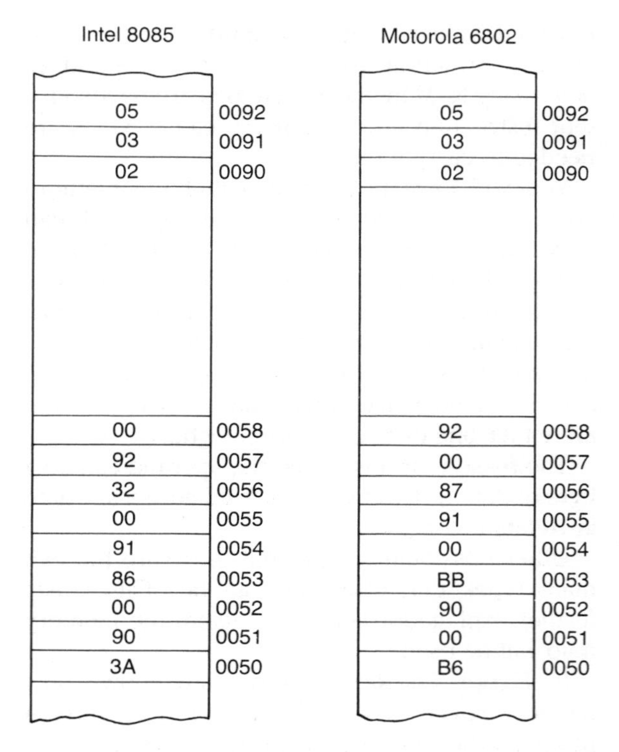

Figure 7.8 Comparison of memory maps for each manufacturer. Notice that the contents are in HEX involving instructions from each instruction set

(a) Load accumulator with the contents of Immediate next memory byte.
(b) Add accumulator with the contents of Immediate next memory byte.
(c) Store accumulator contents into Immediate next memory byte – PROBLEM!

We find that neither Intel or Motorola will allow us to store accumulator contents in this way! The first two instructions can be carried out in the Immediate mode, but we have to use the same addressing mode as previously to store accumulator contents.

The Immediate mode will use different op-codes but the programme and its data will now look like this:

Address	0060	Load accumulator with
	0061	Data in 0061
	0062	Add to accumulator the
	0063	Data in 0063
	0064	Store accumulator contents

0065 into address 0092
0066

Using the Immediate op-codes for both Intel and Motorola processors gives the following programs as shown in Figure 7.9.

	Intel	Motorola
Load accumulator (IMM)	3E 02	86 02
Add to accumulator (IMM)	C6 03	8B 03
Store accumulator into address 0092	32 92 00	B7 00 92

Figure 7.9 How the IMMEDIATE mode involves the data needed so that it follows IMMEDIATELY after the instruction

The memory map for both microprocessors after the program has been run would look like those in Figure 7.10.

The two addressing modes described above, Extended and Immediate for Motorola, Direct and Immediate for Intel, can be used on very many occasions and have been used here to convey the concepts of memory map, addressing modes, operations, operands and op-codes. The op-codes that have been used in the simple example program have been taken from the data sheets of those manufacturers.

A full listing of the instruction sets is given in Appendix A.

Some of the other instruction types include operations such as Subtract, Logical AND, OR, Shift, Compare, Increment and Decrement. These are typical instructions used in many programs.

You have seen two and three-byte instructions which use an op-code plus either one or two further bytes of memory. Some of the single-byte instructions will be mentioned now so that you can see that all is contained in the instruction and no further operands are necessary.

Motorola:

- CLR CLEAR ACCUMULATOR – set contents to 00
- COMA COMPLEMENT ACCUMULATOR – flip bits 0–1; 1–0
- DECA DECREMENT ACCUMULATOR – reduce value of contents of accumulator by 1.

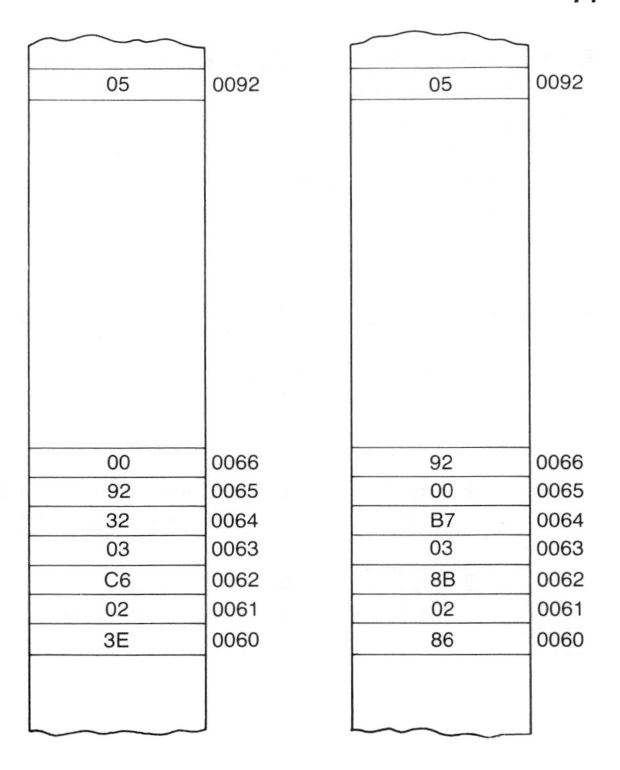

Figure 7.10 Memory map for each manufacturer, showing HEX instructions for the program in machine code for Figure 7.9

Intel:

- CMA COMPLEMENT ACCUMULATOR – same as above
- NOP NO OPERATION – a 'do nothing' state
- HLT HALT – stop processor action.

Summary

In this chapter you have met the memory chips, the ROM and RAM, that make a viable system. The terms *hardware*, *firmware* and *software* have been explained and the importance and use of each type of chip has been developed. The concept of a *memory map* for the microprocessor, into which we can place our memory chips where we want to, has been learned.

It has been seen that RAM chips are used for writing and developing programs before storing permanently, either on disk or in EPROM, and that

RAM is *usually* found in the lower parts of the memory map address range.

Finally, the *op-codes* of two microprocessors were very briefly introduced to show how the microprocessor takes our programming concepts and would recognise in each op-code all the information needed to execute that particular instruction.

The idea of *addressing modes* has been covered at a simple level to show that where the data is placed determines the address mode used.

Suggested Assignment Work

1 The Motorola 6802 microprocessor has other addressing modes such as Direct and Indexed. Discover from the instruction sets in the appendices the meaning of these terms.

2 The Intel 8085 microprocessor has other addressing modes such as Register and Register indirect. Once again find out the meaning of these terms.

3 Explain why Immediate instructions need less memory and less execution time for both microprocessors than the Extended mode.

EIGHT

Programming a microprocessor-based system

OBJECTIVES

At the end of this unit all students should be able to:

Deduce a flow diagram and program sequence for a simple process, e.g. domestic water heating system and possible other systems such as:
a compact disc player
an automatic suspension levelling system.

All the component parts of our microelectronic systems have been described and examined and it would be useful to look back at the past seven chapters to review the various stages covered before re-examining some of the systems that were met in the first chapter.

Chapter Two reviews the symbols that are in common use in flow charts and presents flow charts showing typical processes that could take place in a washing machine, a central heating system, a microwave oven, a cassette or compact disc player and an active suspension system.

Chapter Three examines the differences between analogue and digital systems and also mentions some of the techniques used in conversion from one form of signal to the other.

Chapters Four and Five describe the microprocessor as a digital system, examining some of the construction stages of a microelectronic active device and the components of a microprocessor needed to build up a complete microprocessor based system.

Chapter Six describes the various peripheral devices that allow the user to interface with the system, to use it and to store programs and data.

Chapter Seven takes this further, discussing the various levels of terminology such as hardware, software and firmware and the parts that relate to each type of term.

In this final chapter all the ingredients or components are now to hand to enable us to take the simple flow charts in Chapter Two and use them to suggest how a system designer would interpret them in the production of programs and data which would make a system work properly and efficiently. What basically has to happen here is the transformation of the flow chart of the required system into the programs which could be stored as firmware in ROM and then used in a system to operate it.

The two microprocessors used as reference throughout this book, the Motorola 6802 and the Intel 8085, will have their instruction sets used in a relatively simple way to show how the flow charts could be used to develop the necessary mnemonic instructions and the further process of converting that into machine recognisable code in hexadecimal format.

As a particular example Figure 8.1 shows a possible flow chart for an imaginary microprocessor-controlled compact disc player, first seen in Chapter Two (page 15).

You will see there are five main process boxes

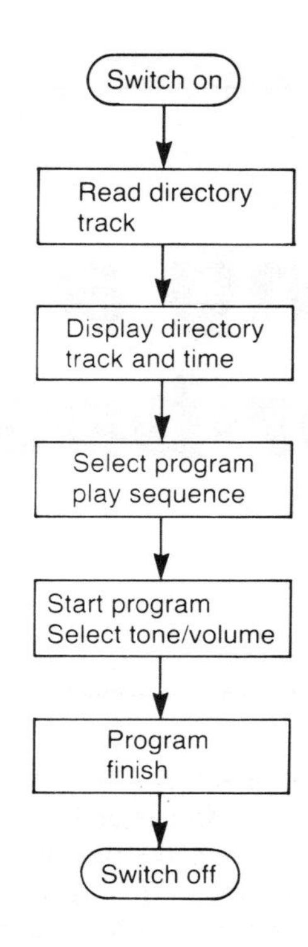

Figure 8.1 Flow chart to operate a compact disc player

each of which needs converting into the necessary instructions which will allow the microprocessor to control the compact disc player so that it plays the tracks on the disc in the order desired by the listener. There are normally three modes available:

(a) Normal play – all tracks are played from start to finish
(b) Programmed play – numbered tracks are selected in a particular sequence and then played.
(c) Shuffle play – all tracks are played but in a random sequence.

Each mode of play has to be handled differently but the main process is still the same. Each track has to be identified, selected and played.

We will now look at each process block and suggest a possible sequence of program instructions to make that block work properly.

On the directory track on the compact disc is stored information about the total number of tracks and the total playing time available on that

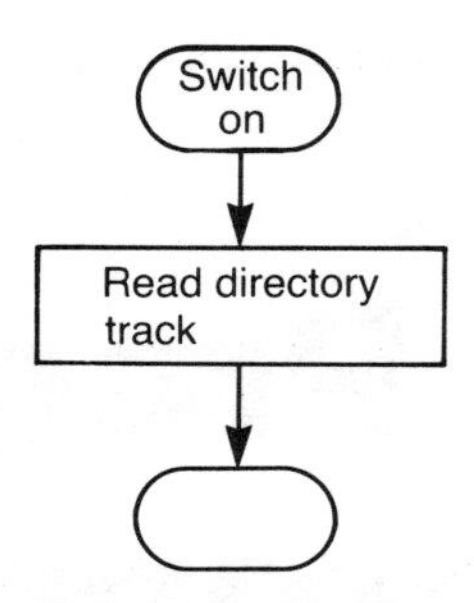

Figure 8.2 First process block to read the directory track

disc. Let us assume that this information has to be read from the disc as it rotates and stored temporarily in the RAM memory of the microprocessor. This has to be done by using the accumulator of the microprocessor and knowing how the microprocessor itself handles input/output processes.

Let us try to describe the process as:

First Block

Read directory track

In Figure 8.2 the process block shows 'Read directory track'. As a result of this process the system will need to know the number of tracks and the total playing time. The microprocessor controlling the system will allocate storage in RAM to hold this information once the directory track is read.

Load accumulator with number of tracks
Store data in RAM memory
Load accumulator with total playing time
Store data in RAM memory

An obvious question now arises – where in RAM memory will this information be stored? This has to be decided by the system designer but we will allocate two fictitious hexadecimal addresses:

0100 for number of tracks
0110 for total playing time

to store this data.

We will assume for the moment that we do not need to worry how the microprocessor handles actual input/output processes which will simplify the program:

Load accumulator with N (number of tracks)

Store accumulator into 0100 (hexadecimal address)
Load accumulator with *T* (total playing time)
Store accumulator into 0110 (hexadecimal address)

In mnemonic form, this could be written:

LDA N
STA 0100
LDA T
STA 0110

Where we place this program in the microprocessor RAM area is up to us to decide, but for the sake of this example let us assume the program starts at address 0200, and also that we can find the value of *N* stored in location 0700 and *T* stored in location 0701.

If the Motorola and Intel mnemonics and op-codes are used, this simple process block could then be rewritten like this:

(a) **Motorola**

Address	*Mnemonics*	*Comments*
0200	LDA A	Load accumulator A (Motorola 6802 has two accumulators called A and B) from address 0700 hex
0201	07	
0202	00	
0203	STA A	Store accumulator A contents into address 0100 hex
0204	01	
0205	00	
0206	LDA A	Load accumulator A from address 0701 hex
0207	07	
0208	01	
0209	STA A	Store accumulator A contents into address 0110 hex
020A	01	
020B	10	

Using the Motorola data sheets to be found in Appendix A and restricting ourselves to the Extended mode of addressing, the above mnemonic program can be written like this:

Address	*Op-code and operands*	*Comments*
0200	B6	
0201	07	
0202	00	
0203	B7	
0204	01	
0205	00	
0206	B6	
0207	07	
0208	01	
0209	B7	
020A	01	
020B	10	
RAM addresses	0100	Temporary store for value of *N*
	0110	Temporary store for value of *T*
	0700	Data read in (*N*)
	0701	Data read in (*T*)

This *machine code* program as we call the combined op-code and operand part of the table, could be written into a ROM structure inside the compact disc player. A certain amount of RAM would be needed at the addresses 0100, 0110, 0700, 0701 as suggested.

(b) **Intel**

Address	*Mnemonics*	*Comments*
0200	LDA	Load accumulator with the contents of hex address 0700 (note the way the address is written!)
0201	00	
0202	07	
0203	STA	Store accumulator content into hex address 0100
0204	00	
0205	01	
0206	LDA	Load accumulator with the contents of hex address 0701
0207	01	
0208	07	
0209	STA	Store accumulator contents into hex address 0110
020A	10	
020B	01	

Again, using the Intel data sheets found in Appendix A, the above mnemonic program can be written as follows:

Address	*Op-codes and operands*
0200	3A
0201	00
0202	07
0203	32
0204	00
0205	01
0206	3A
0207	01
0208	07
0209	32
020A	10
020B	01

The RAM addresses needed would be exactly the same as in the Motorola example.

This is the process which has to be carried out for each of the various process blocks in turn. It has been very greatly simplified for the purposes of explanation but, in principle, it is what is done at introductory levels in microelectronics courses.

There are very many ways of making this process more efficient such as using devices called *assemblers*. These do the actual conversion of the mnemonics directly into op-codes and operands. *Cross-assemblers* allow the same process to be carried out on other computers. *High-level languages* such as PASCAL, C and MODULA-II use compilers to convert 'English-style' statements that we find easy to understand into the machine code easily understood by the microprocessor itself.

Second Block

It is one thing to read the directory track *but* the user *needs to know* by looking at the visual display *how many* tracks there are and *how long* the playing time is. This is very nearly a repeat of the first block except that, instead of reading the information in, it will be written out to the display panel.

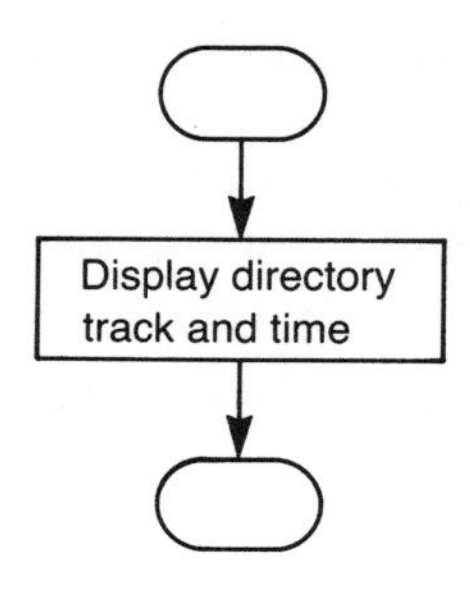

Figure 8.3 Second process block to display the number of tracks and playing time

Figure 8.3 shows the process to display the number of tracks and playing time. The microprocessor organises this by allocating further locations to store this information.

For the sake of simplicity, we will allocate two more RAM addresses to handle the output display, which would be temporary locations to store the data concerning the total number of tracks and total playing time.

The Input/Output Process

The two microprocessors have different ways of handling input/output functions.

The Motorola 6802 uses interface chips such as a Programmable Interface Adapter (PIA) to interface the data bus to the outside world. To do this, a few locations in the memory map are assigned to registers in the PIA. When the microprocessor accesses those addresses it does not have to know that what is really happening is that the pattern on the data bus is being passed, via the PIA, to the outside world.

The Intel 8085 uses separate memory input/output instructions to separate the flow of data between the MPU and memory and from the MPU to the outside world. The Intel microprocessor used instructions like

```
IN     XX
OUT    YY
```

where XX and YY represent hexadecimal port numbers. Connected to these ports would be the interface chips such as the PPI (Programme

Peripheral Interface) or PIO (Programmable Input/Output).

Let us assume that two further RAM locations 0702 and 0703 are made available to show the values of *T* and *N* on the output display.

Note The values of *T* and *N* will be binary numbers and will need additional processing to allow the display to show data which is meaningful to us. For the sake of simplicity, however, we will assume that this has been done. The microprocessor will have to copy the data in locations 0700 and 0701 into 0702 and 0703 which will then be sent out to the display. The simplified program could look something like this:

Load accumulator with *N*
Store accumulator into 0702 (hexadecimal address)
Write data to output display
Load accumulator with *T*
Store accumulator into 0703 (hexadecimal address)
Write data to output display

In mnemonic form this could be written:

LDA *N*
STA 0702
STA OUTPUT DISP 1 – Display of number of track
LDA *T*
STA 0703
STA OUTPUT DISP 2 – Display of total playing time

Because of the different ways in which the microprocessors handle input/output processes, the programs for each microprocessor will differ slightly, as shown below.

We will assume that this part of the program starts at RAM location 0210.

(a) **Motorola**

Address	*Mnemonics*	*Comments*
0210	LDA A	Load accumulator A with contents of address 0100 (= *N*)
0211	01	
0212	00	
0213	STA A	Store contents of accumulator A into address 0702 causing data to go directly to output display
0214	07	
0215	02	
0216	LDA A	Load accumulator A with contents of address 0110 (= *T*)
0217	01	
0218	10	
0219	STA A	Display contents of accumulator A into address 0703 causing data to go directly to display
021A	07	
021B	03	

In a similar way to the first block ('Read Directory Track') the second block is converted into machine code as:

Address	*Op-codes and operands*
0210	B6
0211	01
0212	00
0213	B7
0214	07
0215	02
0216	B6
0217	01
0218	10
0219	B7
021A	07
021B	03

Additional RAM addresses: 0702 Write data out (*N*); 0703 Write data out (*T*)

(b) **Intel**

In this case, the program will be slightly different because of using the Out instructions.

Address	*Mnemonics*	*Comments*
0210	LDA	Load accumulator with contents of address 0100 (= *N*)
0211	00	
0212	01	

Address	*Mnemonics*	*Comments*
0213	OUT	Send this to output port 20
0214	20	(assumed to be part of display)
0215	LDA	Load accumulator with contents of address 0110 (= *T*)
0216	01	
0217	10	
0218	OUT	Send this to output port 25
0219	25	(assumed to be part of display)

What has happened here is that output ports 20 and 25 have been assigned to the display. The number of memory locations in RAM used by this program is two fewer than that used by the Motorola version.

This program could be converted into Intel op-codes as shown below:

Address	*Op-codes and operands*
0210	3A
0211	00
0212	01
0213	D3
0214	20
0215	3A
0216	01
0217	10
0218	03
0219	25

Third Block

It was said earlier that there could be three modes of play:

(a) play each track in order from the first to the last
(b) play each track in a pre-programmed sequence
(c) play tracks in a random manner decided by the CD player itself.

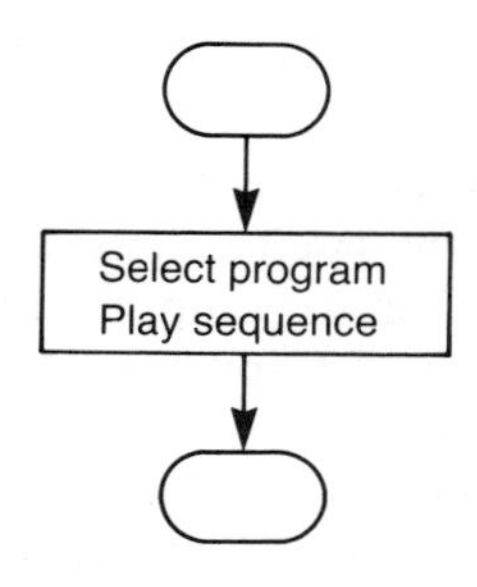

Figure 8.4 Third process block to control the program play sequence

Figure 8.4 shows the process by which the user controls the order of playing the tracks on the disc. The microprocessor will need to organise additional storage in RAM to store the order of playing.

We will only consider case (b) here where *we* can choose the order of playing each track. What could happen here is that when we press the 'Program Play' button on the front of the CD player, a routine is set up which allocates a number of temporary storage areas in RAM, *N* in number, and we put into each location in turn the *number* of the track that we wish to play and this controls the sequence.

Let us assume that $N = 10$ tracks, the number stored in 0100. The program has to allocate 10 separate RAM locations and we will assume that they start at 0704. Our program will ask us to choose each track and store it in following RAM addresses. Suppose the sequence was chosen to be something like this:

First choice	0704
Second choice	0705
Third choice	0706
Ninth choice	070C
Tenth choice	070D hexadecimal address

In this case we would need to know when we have finished choosing the order. Figure 8.5, which was first seen in Chapter Two, page 15, shows this.

The program could be described as shown:

Load accumulator with number of tracks
Store it in memory 0101
Choose number of a track to play

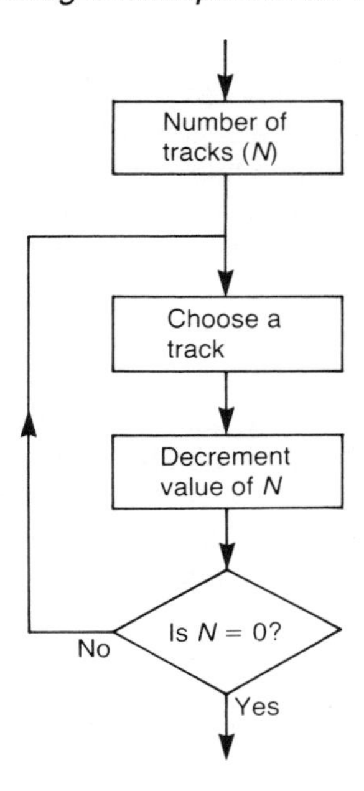

Figure 8.5 Flow chart to show choice of playing order

Store it in 0704
Increment memory pointer
Decrement contents of 0101
Is content of 0101 equal to zero?
If No jump back to 'choice of track' and continue
If Yes – Finish.

This could be represented by:

```
        LDA N
        STA 0101
LOOP    READ IN DATA (choice of track)
        STA 0704
        INC Memory pointer
        DEC 0101
        BRANCH BACK TO LOOP IF 0101
          CONTENTS = 0
        END
```

At this point the way each manufacturer approaches this task of modifying a memory pointer and branching are different and need a certain amount of explanation. We will consider each task in turn.

Memory pointers Intel uses a pair of registers called the HL register pair to access memory. For example, whenever one of their instructions is used that moves a byte between the accumulator and memory, such as:

MOV A,M (moves a byte from memory into the accumulator)
or MOV M,A (moves a byte from the accumulator into memory required)

the system knows automatically that it can find the address of memory. It does this by examining the contents of the HL register pair, acting together. The contents of these two registers form the full 16-bit address to which the system will go, or access, to complete the move instruction. By incrementing or decrementing the contents of HL a series of memory locations can be accessed, one after the other.

For example, LXI H, 3412 has the effect of loading HL pair with the address 1234.
MOV A,M has the effect of moving *to* the accumulator A from memory the contents of the address 1234.

Motorola uses a mode of addressing called the *indexed* addressing mode, which uses a 16-bit register called the *index register*. Then, by incrementing or decrementing the contents of the index register, a series of sequential memory locations can be accessed.
For example,

LDX 1234 has the effect of loading the index register with the address 1234
LDA O,X has the effect of loading the accumulator with the contents of the index register plus zero.
The contents of the index register may be incremented or decremented.

So, each manufacturer provides a similar function, but by rather different means.

Branching This is a very powerful tool to have available. You may remember that whenever the ALU carries out an instruction, special events can affect the flags register. These special events include the following conditions:

Has a calculation just completed resulted in a negative number?
Has an addition just completed generated a carry bit?
Has a calculation just completed resulted in zero?

The flags register contains a number of bits which individually and continually monitor a number of special conditions. These include the three quoted above.

We can make the microprocessor repeat a series of instructions a number of times by putting a number into a register, processing some data, decrementing the register and *testing* if its contents are zero.

If they are zero, the program will follow one path, if *not zero* it can be made to *branch* back to an earlier point and repeat the sequence.

For example, in the INTEL processor, a small segment of a program could look like this:

```
LOOP1  MVI B 04
       NOP
       NOP
       NOP
       DCR B
       JNZ LOOP1
```

This trivial little program loads 4 into register B, does *nothing* for the next three instructions, then decrements B so that it reduces from 4 to 3 and then tests the zero flag.

Is it zero? Answer – No, not yet!

So, the program returns to LOOP1 (we call this a label) and continues and repeats until the content of B *is* zero.

The Motorola processor may well have a similar program segment like this:

```
LOOP1  LDA B 04
       NOP
       NOP
       NOP
       DEC B
       BNE LOOP1
       SWI
```

In this case, the mnemonics are different *but* the effect is just the same! After four passes through the program, the contents of the second accumulator B are progressively reduced from 4 down to zero, when the program segment finishes.

We can now continue these techniques to allow each processor to accept the string of channel numbers that we have chosen to load into our CD player.

Intel routine

	LDA *N*	Load into accumulator number of tracks
	MOV B, A	Move *N* into register B
	LXI H 0407	Load HL with 0704 address
LOOP	IN (Port No)	Read in a number of our choice on a particular port
	MOV M, A	Move contents of A to memory (address = 0704)
	INX H	Increment HL pair
	DCR B	Decrement register B contents
	JNZ LOOP	Is content of B = 0 (Have *we* completed *our* selection?)
	HLT	Program finishes

Motorola routine

	LDA B N	Load accumulator B with *N*
	STA B 0101	Store *N* into second accumulator B
	LDX 0704	Load index register with 0704
LOOP	LDA A DATA	Choice of track number is on data
	STA A O, X	Store data into 0704
	INX	Increment index register to 0705
	DEC 0101	Decrement contents of 0101 from *N* to *N* – 1
	BNE LOOP	Test to see if content of 0101 = 0
	SWI	Program finishes

So, these routines for either processor will allow us to enter our choice of track play sequence into a number of memory locations which can then be used in the next block when the tracks are actually played.

Fourth Block

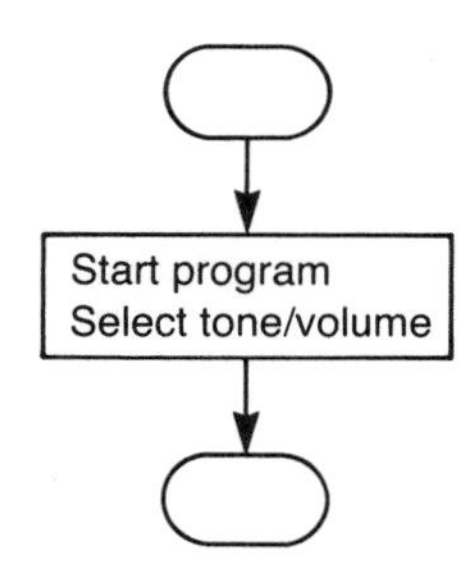

Figure 8.6 Fourth process block showing how the user sets the required tone and volume controls

In this block, shown in Figure 8.6 above, 'select tone and volume' is a process best left to the listener and will not be controlled by the program.

The Start Program sequence will be basically a repeat of block 3 involving a loopcounter which will choose each memory location in turn and play the track number that is stored there. You could, as an exercise on your own, try to devise a routine to do this.

So, our compact disc player can be programmed to carry out a series of tasks that *we* choose to give it.

This process can be repeated in many different ways and for other very different tasks, some of which have already been mentioned earlier, such as:

a domestic central heating system
an active suspension system for a racing car.

Because there are one or two special aspects of those two tasks we will consider them to see how they affect the writing of a program.

DOMESTIC CENTRAL HEATING SYSTEM

In Chapter Two a flow chart was shown of the possible sequences that a central heating system could be expected to go through. The diagram is repeated in Figure 8.7.

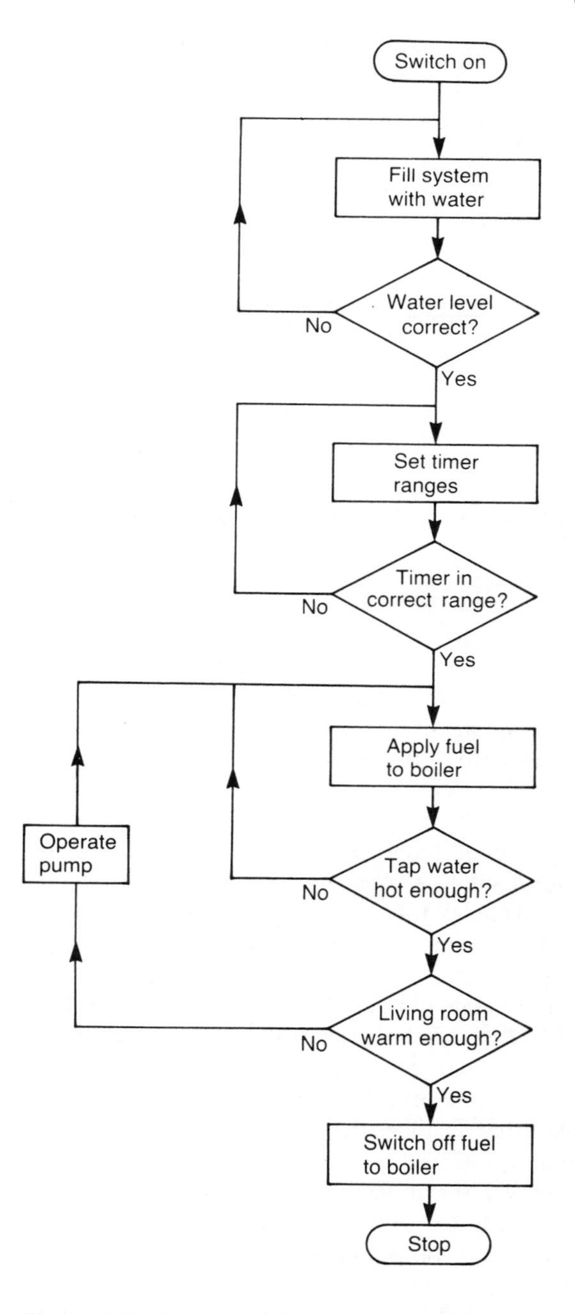

Figure 8.7 Flow chart for a central heating system

In the first block the system water level has to be checked on intial fill-up but also as a safety measure.

The decision box relies on a transducer in the header tank, or other means of supply, to the

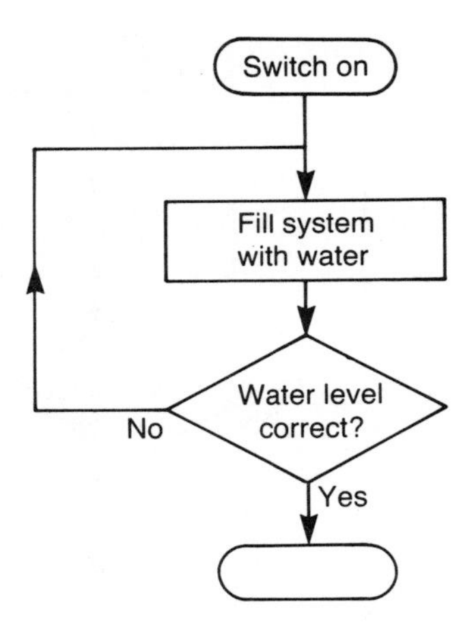

Figure 8.8 Part of the flow chart to check water level

heating system. It could be a linear but analogue level detector or a simple on/off device which tells the system that the minimum water level is reached. Part of this process could be interrogating the state of the switch until it closes when the water level is correct.

The blocks involving the timers are once again simple on/off timers which control fuel flow to the boiler itself; this does not need any form of digital control except that some boiler manufacturers are providing digital timers instead of analogue ones.

The blocks containing questions on the temperature state of the hot water supplied to the taps and to the radiators raises the question of *priorities*. The decision blocks as shown in Figure 8.9 indicate that priority is to be given to the hot water that is to go to the taps, rather than the radiators.

Hence a thermostat connected to the hot water tank will have priority over the room thermostat in the living room and our controller will be programmed to check the temperature of the hot water first before switching on the central heating pump which would then circulate hot water to the radiators.

Although not shown in Figure 8.7 the process would not start and stop as shown but could be connected in a continuous loop *or* only connected in a continuous loop once the timers are operating.

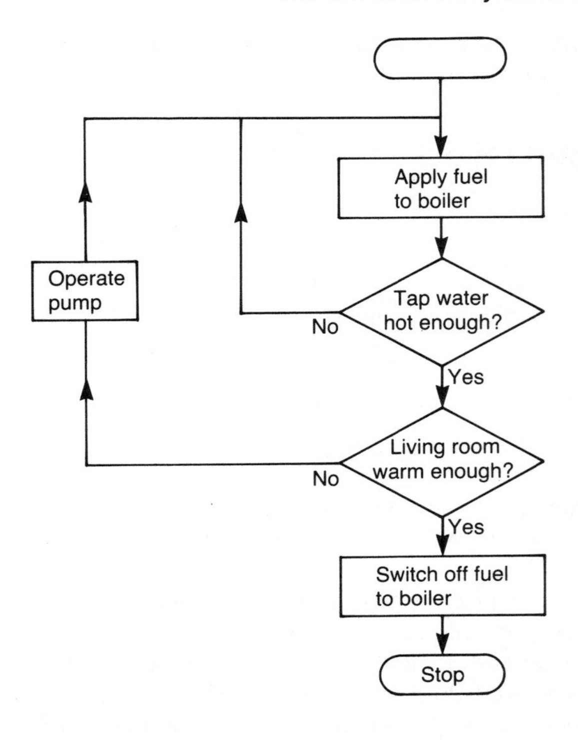

Figure 8.9 Part of the flow chart to set required temperatures of water and rooms

ACTIVE SUSPENSION SYSTEM FOR A RACING CAR

The flow chart, first shown in Chapter Two, is shown again opposite in Figure 8.10.

Unlike the domestic central heating system just described, this system is highly dynamic and the processing speed has to be high to cope with the cornering speeds of a modern Grand Prix racing car.

As each wheel is rotating it experiences both vertical reaction and horizontal sliding forces and the body of the car will move through an angle called the roll angle. There have to be sensors which are capable of accurate measurement in real-time over a range of movements which then should be converted into a digital signal that the microprocessor can handle. This obviously calls for very fast analogue-to-digital conversion (ADC) such as the 'flash' mentioned in an earlier chapter.

Each wheel will have to have its own set of sensors and the processor will have to be fast

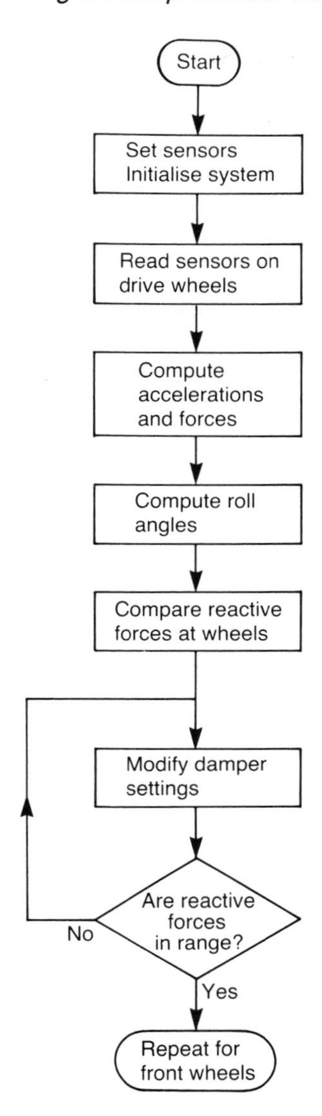

Figure 8.10 Flow chart showing operation of an active suspension system

enough to react to rapid changes in vehicle attitude.

The microprocessor will also have to compute the necessary corrections and modifications to the variable damper settings on each of the wheels and, of course, this will need rapid and accurate digital-to-analogue conversion if the damper rates are to be correctly set within the range available.

This process should be repeated throughout the Grand Prix to maintain the attitude of the racing car at all times. Obviously external factors such as aerodynamic lift if the spoilers are incorrectly set and the risk of aquaplaning in torrential rain will also be inputs to this system.

This software will have to be fast and agile in real time if the system is to be successful. This is a combination of efficient real-time programming and fast computer power.

Programming the microprocessor itself is a complex task and it would be foolish to pretend that the simple examples of programs in this book would be used in a real situation. The point we have tried to make is that, with a thorough knowledge of the instruction set of a chosen microprocessor, experience in devising algorithms for solution which work well enough, fast enough and *safely* enough can lead to efficient machine code programs for a given application.

It is left to other books dealing with this subject at higher levels to consider the above points in finer detail.

Summary

This chapter has brought together many of the topics and ideas of earlier chapters in order to act as a framework for some highly simplified examples of programs for microprocessor applications

The two microprocessors chosen, the Intel 8085 and the Motorola 6802, have been used extensively in schools and technical colleges all over the country in courses up to National Certificate level.

The intention for this introductory level has been to examine a limited range of possibilities available from the instruction set for each machine. It is left to the reader to devise, experiment, and modify his or her own programs to carry out simple tasks.

Suggested Assignment Work

1 Suggest sensors for the following appliances:

(a) talking dashboard display

(b) automatic focussing camera

(c) microwave oven.

2 Devise possible programs which could be implemented by the Intel 8085 or Motorola 6802 microprocessors for the appliance in 1.

Appendix A

Instruction sets for Motorola and Intel microprocessors

The copyrights of both Intel and Motorola and their permission to publish extracts from their data sheets are gratefully acknowledged.

Motorola
6802 Table 3: accumulator and memory instructions
6802 Table 4: index register and stack manipulation instructions
6802 Table 5: jump and branch instructions

Intel
8085 Instruction set index
8085 Numeric summary of instructions

Motorola 6802 Table 3 – Accumulator and memory instructions

OPERATIONS	MNEMONIC	ADDRESSING MODES IMMED			DIRECT			INDEX			EXTND			IMPLIED			BOOLEAN/ARITHMETIC OPERATION (All register labels refer to contents)	COND. CODE REG. 5	4	3	2	1	0
		OP	~	#	OP	~	#	OP	~	#	OP	~	#	OP	~	#		H	I	N	Z	V	C
Add	ADDA	8B	2	2	9B	3	2	AB	5	2	BB	4	3				A + M → A	↕	•	↕	↕	↕	↕
	ADDB	CB	2	2	DB	3	2	EB	5	2	FB	4	3				B + M → B	↕	•	↕	↕	↕	↕
Add Acmltrs	ABA													1B	2	1	A + B → A	↕	•	↕	↕	↕	↕
Add with Carry	ADCA	89	2	2	99	3	2	A9	5	2	B9	4	3				A + M + C → A	↕	•	↕	↕	↕	↕
	ADCB	C9	2	2	D9	3	2	E9	5	2	F9	4	3				B + M + C → B	↕	•	↕	↕	↕	↕
And	ANDA	84	2	2	94	3	2	A4	5	2	B4	4	3				A · M → A	•	•	↕	↕	R	•
	ANDB	C4	2	2	D4	3	2	E4	5	2	F4	4	3				B · M → B	•	•	↕	↕	R	•
Bit Test	BITA	85	2	2	95	3	2	A5	5	2	B5	4	3				A · M	•	•	↕	↕	R	•
	BITB	C5	2	2	D5	3	2	E5	5	2	F5	4	3				B · M	•	•	↕	↕	R	•
Clear	CLR							6F	7	2	7F	6	3				00 → M	•	•	R	S	R	R
	CLRA													4F	2	1	00 → A	•	•	R	S	R	R
	CLRB													5F	2	1	00 → B	•	•	R	S	R	R
Compare	CMPA	81	2	2	91	3	2	A1	5	2	B1	4	3				A – M	•	•	↕	↕	↕	↕
	CMPB	C1	2	2	D1	3	2	E1	5	2	F1	4	3				B – M	•	•	↕	↕	↕	↕
Compare Acmltrs	CBA													11	2	1	A – B	•	•	↕	↕	↕	↕
Complement, 1's	COM							63	7	2	73	6	3				$\overline{M}$ → M	•	•	↕	↕	R	S
	COMA													43	2	1	$\overline{A}$ → A	•	•	↕	↕	R	S
	COMB													53	2	1	$\overline{B}$ → B	•	•	↕	↕	R	S
Complement, 2's	NEG							60	7	2	70	6	3				00 – M → M	•	•	↕	↕	(1)	(2)
(Negate)	NEGA													40	2	1	00 – A → A	•	•	↕	↕	(1)	(2)
	NEGB													50	2	1	00 – B → B	•	•	↕	↕	(1)	(2)
Decimal Adjust, A	DAA													19	2	1	Converts Binary Add. of BCD Characters into BCD Format	•	•	↕	↕	↕	(3)
Decrement	DEC							6A	7	2	7A	6	3				M – 1 → M	•	•	↕	↕	4	•
	DECA													4A	2	1	A – 1 → A	•	•	↕	↕	4	•
	DECB													5A	2	1	B – 1 → B	•	•	↕	↕	4	•
Exclusive OR	EORA	88	2	2	98	3	2	A8	5	2	B8	4	3				A ⊕ M → A	•	•	↕	↕	R	•
	EORB	C8	2	2	D8	3	2	E8	5	2	F8	4	3				B ⊕ M → B	•	•	↕	↕	R	•
Increment	INC							6C	7	2	7C	6	3				M + 1 → M	•	•	↕	↕	(5)	•
	INCA													4C	2	1	A + 1 → A	•	•	↕	↕	(5)	•
	INCB													5C	2	1	B + 1 → B	•	•	↕	↕	(5)	•
Load Acmltr	LDAA	86	2	2	96	3	2	A6	5	2	B6	4	3				M → A	•	•	↕	↕	R	•
	LDAB	C6	2	2	D6	3	2	E6	5	2	F6	4	3				M → B	•	•	↕	↕	R	•
Or, Inclusive	ORAA	8A	2	2	9A	3	2	AA	5	2	BA	4	3				A + M → A	•	•	↕	↕	R	•
	ORAB	CA	2	2	DA	3	2	EA	5	2	FA	4	3				B + M → B	•	•	↕	↕	R	•
Push Data	PSHA													36	4	1	A → M_{SP}, SP – 1 → SP	•	•	•	•	•	•
	PSHB													37	4	1	B → M_{SP}, SP – 1 → SP	•	•	•	•	•	•

Operation	Mnemonic	IMMED	DIRECT	INDEX	EXTND	IMPLIED	Boolean/Arithmetic	H	I	N	Z	V	C
Pull Data	PULA					32 4 1	SP + 1 → SP, M_{SP} → A	•	•	•	•	•	•
	PULB					33 4 1	SP + 1 → SP, M_{SP} → B	•	•	•	•	•	•
Rotate Left	ROL			69 7 2	79 6 3		M	•	•	↕	↕	(6)	↕
	ROLA					49 2 1	A } C ← b7 ← b0	•	•	↕	↕	(6)	↕
	ROLB					59 2 1	B	•	•	↕	↕	(6)	↕
Rotate Right	ROR			66 7 2	76 6 3		M	•	•	↕	↕	(6)	↕
	RORA					46 2 1	A } C → b7 → b0	•	•	↕	↕	(6)	↕
	RORB					56 2 1	B	•	•	↕	↕	(6)	↕
Shift Left, Arithmetic	ASL			68 7 2	78 6 3		M	•	•	↕	↕	(6)	↕
	ASLA					48 2 1	A } C ← b7 ← b0 ← 0	•	•	↕	↕	(6)	↕
	ASLB					58 2 1	B	•	•	↕	↕	(6)	↕
Shift Right, Arithmetic	ASR			67 7 2	77 6 3		M	•	•	↕	↕	(6)	↕
	ASRA					47 2 1	A } b7 → b0 → C	•	•	↕	↕	(6)	↕
	ASRB					57 2 1	B	•	•	↕	↕	(6)	↕
Shift Right, Logic	LSR			64 7 2	74 6 3		M	•	•	R	↕	(6)	↕
	LSRA					44 2 1	A } 0 → b7 → b0 → C	•	•	R	↕	(6)	↕
	LSRB					54 2 1	B	•	•	R	↕	(6)	↕
Store Acmltr	STAA		97 4 2	A7 6 2	B7 5 3		A → M	•	•	↕	↕	R	•
	STAB		D7 4 2	E7 6 2	F7 5 3		B → M	•	•	↕	↕	R	•
Subtract	SUBA	80 2 2	90 3 2	A0 5 2	B0 4 3		A − M → A	•	•	↕	↕	↕	↕
	SUBB	C0 2 2	D0 3 2	E0 5 2	F0 4 3		B − M → B	•	•	↕	↕	↕	↕
Subtract Acmltrs	SBA					10 2 1	A − B → A	•	•	↕	↕	↕	↕
Subtr. with Carry	SBCA	82 2 2	92 3 2	A2 5 2	B2 4 3		A − M − C → A	•	•	↕	↕	↕	↕
	SBCB	C2 2 2	D2 3 2	E2 5 2	F2 4 3		B − M − C → B	•	•	↕	↕	↕	↕
Transfer Acmltrs	TAB					16 2 1	A → B	•	•	↕	↕	R	•
	TBA					17 2 1	B → A	•	•	↕	↕	R	•
Test, Zero or Minus	TST			6D 7 2	7D 6 3		M − 00	•	•	↕	↕	R	R
	TSTA					4D 2 1	A − 00	•	•	↕	↕	R	R
	TSTB					5D 2 1	B − 00	•	•	↕	↕	R	R

LEGEND:

OP	Operation Code (Hexadecimal)
~	Number of MPU Cycles
#	Number of Program Bytes
+	Arithmetic Plus
−	Arithmetic Minus
·	Boolean AND
M_{SP}	Contents of memory location pointed to be Stack Pointer
+	Boolean Inclusive OR
⊕	Boolean Exclusive OR
$\overline{M}$	Complement of M
→	Transfer Into
0	Bit = Zero
00	Byte = Zero

CONDITION CODE SYMBOLS

H	Half carry from bit 3
I	Interrupt mask
N	Negative (sign bit)
Z	Zero (byte)
V	Overflow, 2's complement
C	Carry from bit 7
R	Reset Always
S	Set Always
↕	Test and set if true, cleared otherwise
•	Not Affected

Note – Accumulator addressing mode instructions are included in the column for IMPLIED addressing

Motorola 6802 Table 4 – Index register and stack manipulation instructions

POINTER OPERATIONS	MNEMONIC	IMMED OP	IMMED ~	IMMED #	DIRECT OP	DIRECT ~	DIRECT #	INDEX OP	INDEX ~	INDEX #	EXTND OP	EXTND ~	EXTND #	IMPLIED OP	IMPLIED ~	IMPLIED #	BOOLEAN/ARITHMETIC OPERATION	5 H	4 I	3 N	2 Z	1 V	0 C
Compare Index Reg	CPX	8C	3	3	9C	4	2	AC	6	2	BC	5	3				$X_H - M, X_L - (M+1)$	•	•	(7)	↕	(8)	•
Decrement Index Reg	DEX													09	4	1	$X - 1 \rightarrow X$	•	•	•	↕	•	•
Decrement Stack Pntr	DES													34	4	1	$SP - 1 \rightarrow SP$	•	•	•	•	•	•
Increment Index Reg	INX													08	4	1	$X + 1 \rightarrow X$	•	•	•	↕	•	•
Increment Stack Pntr	INS													31	4	1	$SP + 1 \rightarrow SP$	•	•	•	•	•	•
Load Index Reg	LDX	CE	3	3	DE	4	2	EE	6	2	FE	5	3				$M \rightarrow X_H, (M+1) \rightarrow X_L$	•	•	(9)	↕	R	•
Load Stack Pntr	LDS	8E	3	3	9E	4	2	AE	6	2	BE	5	3				$M \rightarrow SP_H, (M+1) \rightarrow SP_L$	•	•	(9)	↕	R	•
Store Index Reg	STX				DF	5	2	EF	7	2	FF	6	3				$X_H \rightarrow M, X_L \rightarrow (M+1)$	•	•	(9)	↕	R	•
Store Stack Pntr	STS				9F	5	2	AF	7	2	BF	6	3				$SP_H \rightarrow M, SP_L \rightarrow (M+1)$	•	•	(9)	↕	R	•
Indx Reg → Stack Pntr	TXS													35	4	1	$X - 1 \rightarrow SP$	•	•	•	•	•	•
Stack Pntr → Indx Reg	TSX													30	4	1	$SP + 1 \rightarrow X$	•	•	•	•	•	•

COND. CODE REG.

Motorola 6802 Table 5 – Jump and branch instructions

OPERATIONS	MNEMONIC	RELATIVE			INDEX			EXTND			IMPLIED			BRANCH TEST	COND. CODE REG. 5	4	3	2	1	0
		OP	~	#	OP	~	#	OP	~	#	OP	~	#		H	I	N	Z	V	C
Branch Always	BRA	20	4	2										None	•	•	•	•	•	•
Branch If Carry Clear	BCC	24	4	2										C = 0	•	•	•	•	•	•
Branch If Carry Set	BCS	25	4	2										C = 1	•	•	•	•	•	•
Branch If = Zero	BEQ	27	4	2										Z = 1	•	•	•	•	•	•
Branch If ≥ Zero	BGE	2C	4	2										$N \oplus V = 0$	•	•	•	•	•	•
Branch If > Zero	BGT	2E	4	2										$Z + (N \oplus V) = 0$	•	•	•	•	•	•
Branch If Higher	BHI	22	4	2										C + Z = 0	•	•	•	•	•	•
Branch If ≤ Zero	BLE	2F	4	2										$Z + (N \oplus V) = 1$	•	•	•	•	•	•
Branch If Lower or Same	BLS	23	4	2										C + Z = 1	•	•	•	•	•	•
Branch If < Zero	BLT	2D	4	2										$N \oplus V = 1$	•	•	•	•	•	•
Branch If Minus	BMI	2B	4	2										N = 1	•	•	•	•	•	•
Branch If Not Equal Zero	BNE	26	4	2										Z = 0	•	•	•	•	•	•
Branch If Overflow Clear	BVC	28	4	2										V = 0	•	•	•	•	•	•
Branch If Overflow Set	BVS	29	4	2										V = 1	•	•	•	•	•	•
Branch If Plus	BPL	2A	4	2										N = 0	•	•	•	•	•	•
Branch To Subroutine	BSR	8D	8	2										}	•	•	•	•	•	•
Jump	JMP				6E	4	2	7E	3	3				} See Special Operations	•	•	•	•	•	•
Jump To Subroutine	JSR				AD	8	2	BD	9	3				}	•	•	•	•	•	•
No Operation	NOP										01	2	1	Advances Prog. Cntr. Only	•	•	•	•	•	•
Return From Interrupt	RTI										3B	10	1					(10)		
Return From Subroutine	RTS										39	5	1	}	•	•	•	•	•	•
Software Interrupt	SWI										3F	12	1	} See Special Operations	•	•	•	•	•	•
Wait For Interrupt*	WAI										3E	9	1	}	•	(11)	•	•	•	•

* WAI puts Address Bus, R/W, and Data Bus in the three-state mode while VMA is held low.

Intel 8085 Hexadecimal instruction set

JUMP

Hex	Instruction	
C3	JMP	
C2	JNZ	
CA	JZ	
D2	JNC	
DA	JC	
E2	JPO	Adr
EA	JPE	
F2	JP	
FA	JM	
E9	PCHL	

CALL

Hex	Instruction	
CD	CALL	
C4	CNZ	
CC	CZ	
D4	CNC	
DC	CC	Adr
E4	CPO	
EC	CPE	
F4	CP	
FC	CM	

RETURN

Hex	Instruction
C9	RET
C0	RNZ
C8	RZ
D0	RNC
D8	RC
E0	RPO
E8	RPE
F0	RP
F8	RM

RESTART

Hex	Instruction	
C7	RST	0
CF	RST	1
D7	RST	2
DF	RST	3
E7	RST	4
EF	RST	5
F7	RST	6
FF	RST	7

ROTATE†

Hex	Instruction
07	RLC
0F	RRC
17	RAL
1F	RAR

CONTROL

Hex	Instruction
00	NOP
76	HLT
F3	DI
FB	EI

MOVE IMMEDIATE

Hex	Instruction		
06	MVI	B,	
0E	MVI	C,	
16	MVI	D,	
1E	MVI	E,	
26	MVI	H,	D8
2E	MVI	L,	
36	MVI	M,	
3E	MVI	A,	

Acc IMMEDIATE*

Hex	Instruction	
C6	ADI	
CE	ACI	
D6	SUI	
DE	SBI	
E6	ANI	D8
EE	XRI	
F6	ORI	
FE	CPI	

LOAD IMMEDIATE

Hex	Instruction		
01	LXI	B,	
11	LXI	D,	D16
21	LXI	H,	
31	LXI	SP,	

DOUBLE ADD†

Hex	Instruction	
09	DAD	B
19	DAD	D
29	DAD	H
39	DAD	SP

STACK OPS

Hex	Instruction	
C5	PUSH	B
D5	PUSH	D
E5	PUSH	H
F5	PUSH	PSW
C1	POP	B
D1	POP	D
E1	POP	H
F1	POP	PSW*
E3	XTHL	
F9	SPHL	

MOVE

Hex	Instruction	
40	MOV	B,B
41	MOV	B,C
42	MOV	B,D
43	MOV	B,E
44	MOV	B,H
45	MOV	B,L
46	MOV	B,M
47	MOV	B,A
48	MOV	C,B
49	MOV	C,C
4A	MOV	C,D
4B	MOV	C,E
4C	MOV	C,H
4D	MOV	C,L
4E	MOV	C,M
4F	MOV	C,A
50	MOV	D,B
51	MOV	D,C
52	MOV	D,D
53	MOV	D,E
54	MOV	D,H
55	MOV	D,L
56	MOV	D,M
57	MOV	D,A

INCREMENT**

Hex	Instruction	
04	INR	B
0C	INR	C
14	INR	D
1C	INR	E
24	INR	H
2C	INR	L
34	INR	M
3C	INR	A
03	INX	B
13	INX	D
23	INX	H
33	INX	SP

DECREMENT**

Hex	Instruction	
05	DCR	B
0D	DCR	C
15	DCR	D
1D	DCR	E
25	DCR	H
2D	DCR	L
35	DCR	M
3D	DCR	A
0B	DCX	B
1B	DCX	D
2B	DCX	H
3B	DCX	SP

LOAD/STORE

Hex	Instruction	
0A	LDAX	B
1A	LDAX	D
2A	LHDL	Adr
3A	LDA	Adr
02	STAX	B
12	STAX	D
22	SHLD	Adr
32	STA	Adr

SPECIALS

Hex	Instruction
EB	XCHG
27	DAA*
2F	CMA
37	STC†
3F	CMC†

INPUT/OUTPUT

Hex	Instruction	
D3	OUT	D8
DB	IN	

D8 = constant, or logical/arithmetic expression that evaluates to an 8 bit data quantity.

* = all Flags (C, Z, S, P, AC) affected

D16 = constant, or logical/arithmetic expression that evaluates to a 16 bit data quantity.

† = only CARRY affected

Intel 8085 (continued)

MOVE (cont)

Code	Instruction	Operand
58	MOV	E,B
59	MOV	E,C
5A	MOV	E,D
5B	MOV	E,E
5C	MOV	E,H
5D	MOV	E,L
5E	MOV	E,M
5F	MOV	E,A
60	MOV	H,B
61	MOV	H,C
62	MOV	H,D
63	MOV	H,E
64	MOV	H,H
65	MOV	H,L
66	MOV	H,M
67	MOV	H,A
68	MOV	L,B
69	MOV	L,C
6A	MOV	L,D
6B	MOV	L,E
6C	MOV	L,H
6D	MOV	L,L
6E	MOV	L,M
6F	MOV	L,A
70	MOV	M,B
71	MOV	M,C
72	MOV	M,D
73	MOV	M,E
74	MOV	M,H
75	MOV	M,L
	......	
77	MOV	M,A
78	MOV	A,B
79	MOV	A,C
7A	MOV	A,D
7B	MOV	A,E
7C	MOV	A,H
7D	MOV	A,L
7E	MOV	A,M
7F	MOV	A,A

ACCUMULATOR*

Code	Instruction	Operand
80	ADD	B
81	ADD	C
82	ADD	D
83	ADD	E
84	ADD	H
85	ADD	L
86	ADD	M
87	ADD	A
88	ADC	B
89	ADC	C
8A	ADC	D
8B	ADC	E
8C	ADC	H
8D	ADC	L
8E	ADC	M
8F	ADC	A
90	SUB	B
91	SUB	C
92	SUB	D
93	SUB	E
94	SUB	H
95	SUB	L
98	SUB	M
97	SUB	A
98	SBB	B
99	SBB	C
9A	SBB	D
9B	SBB	E
9C	SBB	H
9D	SBB	L
9E	SBB	M
9F	SBB	A
A0	ANA	B
A1	ANA	C
A2	ANA	D
A3	ANA	E
A4	ANA	H
A5	ANA	L
A6	ANA	M
A7	ANA	A
A8	XRA	B
A9	XRA	C
AA	XRA	D
AB	XRA	E
AC	XRA	H
AD	XRA	L
AE	XRA	M
AF	XRA	A
B0	ORA	B
B1	ORA	C
B2	ORA	D
B3	ORA	E
B4	ORA	H
B5	ORA	L
B6	ORA	M
B7	ORA	A
B8	CMP	B
B9	CMP	C
BA	CMP	D
BB	CMP	E
BC	CMP	H
BD	CMP	L
BE	CMP	M
BF	CMP	A

PSEUDO INSTRUCTION

Instruction	Operand	
ORG	Adr	
END		
EQU	D16	
SET	D16	
DS	D16	
DB	D8	[ ]
DW	D16	[ ]
IF	D16	
ENDIF		
MACRO		[ ]
ENDM		

CONSTANT DEFINITION

OBDH 1AH	Hex
105D 105	Decimal
72Q 72O	Octal
1 1 0 1 1 B 0 0 1 1 0 B	Binary
'TEST' 'A' 'B'	ASCII

OPERATORS

(,)
*,/, MOD, SHL, SHR
+, −
NOT
AND
OR,XOR

STANDARD SETS

A	SET	7
B	SET	0
C	SET	1
D	SET	2
E	SET	3
H	SET	4
L	SET	5
M	SET	6
SP	SET	6
PSW	SET	6

FLAG BYTE STACK FORMAT

7	6	5	4	3	2	1	∅
S	Z	∅	AC	∅	P	1	C

Adr – 16 bit address

** – all Flags except CARRY affected; (exception: INX & DCX affect no Flags)

Intel 8085 Numeric summary of instruction set

00	NOP		2B	DCX	H	58	MOV	D,M	81	ADD	C	AC	XRA	H	D7	RST	2
01	LXI	B,D16	2C	INR	L	57	MOV	D,A	82	ADD	D	AD	XRA	L	D8	RC	
02	STAX	B	2D	DCR	L	58	MOV	E,B	83	ADD	E	AE	XRA	M	D9	...	
03	INX	B	2E	MVI	L,D8	59	MOV	E,C	84	ADD	H	AF	XRA	A	DA	JC	Adr
04	INR	B	2F	CMA		5A	MOV	E,D	85	ADD	L	B0	ORA	B	DB	IN	D8
05	DCR	B	30	...		5B	MOV	E,E	86	ADD	M	B1	ORA	C	DC	CC	Adr
06	MVI	B,D8	31	LXI	SP,D16	5C	MOV	E,H	87	ADD	A	B2	ORA	D	DD	...	
07	RLC		32	STA	Adr	5D	MOV	E,L	88	ADC	B	B3	ORA	E	DE	SBI	D8
08	...		33	INX	SP	5E	MOV	E,M	89	ADC	C	B4	ORA	H	DF	RST	3
09	DAD	B	34	INR	M	5F	MOV	E,A	8A	ADC	D	B5	ORA	L	E0	RPO	
OA	LDAX	B	35	DCR	M	60	MOV	H,B	8B	ADC	E	B6	ORA	M	E1	POP	H
OB	DCX	B	36	MVI	M,D8	61	MOV	H,C	8C	ADC	H	B7	ORA	A	E2	JPO	Adr
OC	INR	C	37	STC		62	MOV	H,D	8D	ADC	L	B8	CMP	B	E3	XTHL	
OD	DCR	C	38	...		63	MOV	H,E	8E	ADC	M	B9	CMP	C	E4	CPO	Adr
OE	MVI	C,D8	39	DAD	SP	64	MOV	H,H	8F	ADC	A	BA	CMP	D	E5	PUSH	H
OF	RRC		3A	LDA	Adr	65	MOV	H,L	90	SUB	B	BB	CMP	E	E6	ANI	D8
10	...		38	DCX	SP	66	MOV	H,M	91	SUB	C	BC	CMP	H	E7	RST	4
11	LXI	D,D16	3C	INR	A	67	MOV	H,A	92	SUB	D	BD	CMP	L	E8	RPE	
12	STAX	D	3D	DCR	A	68	MOV	L,B	93	SUB	E	BE	CMP	M	E9	PCHL	
13	INX	D	3E	MVI	A,D8	69	MOV	L,C	94	SUB	H	BF	CMP	A	EA	JPE	Adr
14	INR	D	3F	CMC		6A	MOV	L,D	95	SUB	L	C0	RNZ		EB	XCHG	
15	DCR	D	40	MOV	B,B	6B	MOV	L,E	96	SUB	M	C1	POP	B	EC	CPE	Adr
16	MVI		41	MOV	B,C	6C	MOV	L,H	97	SUB	A	C2	JNZ	Adr	ED	...	
17	RAL		42	MOV	B,D	6D	MOV	L,L	98	SUB	B	C3	JMP	Adr	EE	XRI	D8
18	...		43	MOV	B,E	6E	MOV	L,M	99	SBB	C	C4	CNZ	Adr	EF	RST	5
19	DAD	D	44	MOV	B,H	6F	MOV	L,A	9A	SBB	D	C5	PUSH	B	F0	RP	
1A	LDAX	D	45	MOV	B,L	70	MOV	M,B	9B	SBB	E	C6	ADI	D8	F1	POP	PSW
1B	DCX	D	46	MOV	B,M	71	MOV	M,C	9C	SBB	H	C7	RST	0	F2	JP	Adr
1C	INR	E	47	MOV	B,A	72	MOV	M,D	9D	SBB	L	C8	RZ		F3	DI	
1D	DCR	E	48	MOV	C,B	73	MOV	M,E	9E	SBB	M	C9	RET	Adr	F4	CP	Adr
1E	MVI	E,D8	49	MOV	C,C	74	MOV	M,H	9F	SBB	A	CA	JZ		F5	PUSH	PSW
1F	RAR		4A	MOV	C,D	75	MOV	M,L	A0	ANA	B	CB	...		F6	ORI	D8
20	...		4B	MOV	C,E	76	HLT		A1	ANA	C	CC	CZ		Adr	F7	RST 6
21	LXI	H,D16	4C	MOV	C,H	77	MOV	M,A	A2	ANA	D	CD	CALL		Adr	F8	RM
22	SHLD	Adr	4D	MOV	C,L	78	MOV	A,B	A3	ANA	E	CE	ACI		D8	F9	SPHL
23	INX	H	4E	MOV	C,M	79	MOV	A,C	A4	ANA	H	CF	RST	1	FA	JM	Adr
24	INR	H	4F	MOV	C,A	7A	MOV	A,D	A5	ANA	L	D0	RNC		FB	EI	
25	DCR	H	50	MOV	D,B	7B	MOV	A,E	A6	ANA	M	D1	POP	D	FC	CM	Adr
26	MVI	H,D8	51	MOV	D,C	7C	MOV	A,H	A7	ANA	A	D2	JNC	Adr	FD	...	
27	DAA		52	MOV	D,D	7D	MOV	A,L	A8	XRA	B	D3	OUT	D8	FE	CPI	D8
28	...		53	MOV	D,E	7E	MOV	A,M	A9	XRA	C	D4	CNC	Adr	FF	RST	7
29	DAD	H	54	MOV	D,H	7F	MOV	A,A	AA	XRA	D	D5	PUSH	D			
2A	LHLD	Adr	55	MOV	D,L	80	ADD	B	AB	XRA	E	D6	SUI	D8			

D8 = constant, or logical/arithmetic expression that evaluates to an 8 bit data quantity.

D16 = constant, or logical/arithmetic expression that evaluates to a 16 bit data quantity.

Appendix B

American standard code for information interchange (ASCII)

Bits 4 to 6	—	0	1	2	3	4	5	6	7
	0	NUL	DLE	SP	0	@	P		p
	1	SOH	DC1	!	1	A	Q	a	q
	2	STX	DC2	″	2	B	R	b	r
	3	ETX	DC3	#	3	C	S	c	s
	4	EOT	DC4	$	4	D	T	d	t
	5	ENQ	NAK	%	5	E	U	e	u
Bits 0 to 3	6	ACK	SYN	&	6	F	V	f	v
	7	BEL	ETB	'	7	G	W	g	w
	8	BS	CAN	(	8	H	X	h	x
	9	HT	EM	)	9	I	Y	i	y
	A	LF	SUB	*	:	J	Z	j	z
	B	VT	ESC	+	;	K	[	k	{
	C	FF	FS	,	<	L	/	l	/
	D	CR	GS	–	=	M	]	m	}
	E	SO	RS	.	>	N	↑	n	≈
	F	SI	US	/	?	O	←	o	DEL

Send a 7-bit ASCII character 'H'; even parity, 2 stop bits:

$$H = 48_{16} = 1\,0\,0\,1\,0\,0\,0_2$$

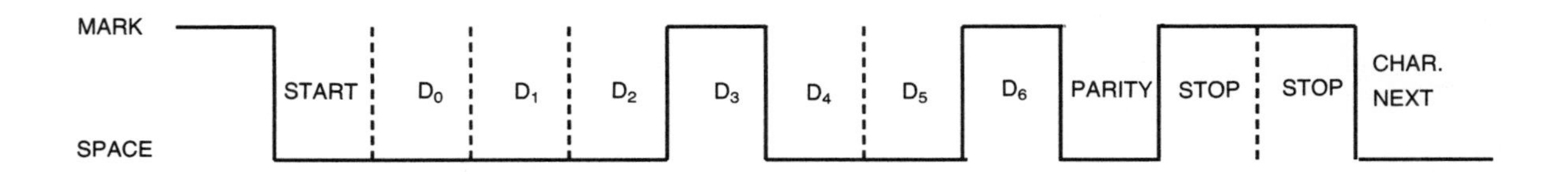

Appendix C

Denary/binary/hexadecimal conversions

Any microprocessor has a language of its own called *machine code*, which humans do not easily learn because it is in binary number code. This means that the language is made up from codes of binary digits (0's and 1's) in groups of eight at a time. (Remember, a group of eight bits is called a *byte*.) For compactness, and the need to be able to represent these codes more efficiently, we use the *hexadecimal* numbering system (base 16, called 'hex' for short) as a form of shorthand. This allows us to replace 8 bits by 2 hex characters. This works on the idea that $16 = 2^4$.

Base 2 has *two* symbols	0 and 1.
Base 10 has *ten* symbols	0, 1, 2, 3, 4, 5, 6, 7, 8 and 9.
Base 16 has *sixteen* symbols	0, 1, 2, 3, 4, 5, 6, 7, 8, 9, and we need *six* more but we cannot allow 10, 11, 12, 13, 14, and 15 because those numbers are *pairs* of digits. So, we invent symbols A, B, C, D, E and F to represent those numbers.

The complete table is shown below.

DENARY	BINARY	HEXADECIMAL
0	0 0 0 0	0
1	0 0 0 1	1
2	0 0 1 0	2
3	0 0 1 1	3
4	0 1 0 0	4
5	0 1 0 1	5
6	0 1 1 0	6
7	0 1 1 1	7
8	1 0 0 0	8
9	1 0 0 1	9
10	1 0 1 0	A
11	1 0 1 1	B
12	1 1 0 0	C
13	1 1 0 1	D
14	1 1 1 0	E
15	1 1 1 1	F

Hence, instead of writing groups of four bits we can write one hex digit, as a form of shorthand. For example, 0 0 1 1 1 1 0 0 can be written as 3C in hex.

Appendix D

Suggested assignment topics

1 Produce block diagrams from other units at first level, e.g. a lathe, a milling machine, a computer numerically controlled (CNC) machine, a robot, an oscilloscope, a power supply unit.

2 Produce and develop flow charts for a piece of electronic test equipment such as a hand-held digital instrument, a modern oscilloscope, a microprocessor based item such as a programmable logic controller (PLC).

3 From your own experience, or by research and reading, suggest a range of transducers to allow operation of an automatic focussing camera, a central heating system, a remote-control device for a television.

4 Seek the views of other students on the problems of safely working with toxic substances producing circuit boards. Discuss or find out about the nature of the various processes involved, the problems of power dissipation and how testing could be carried out.

5 From the data sheets, detail the main differences in processor architecture, register arrays, input/output and control signals between the two processors used, the Motorola 6802 and Intel 8085.

6 Arrange visits to various places which use computer installations and look at the way various input/output media are used in different applications.

7 Analyse the functions expected from a typical monitor program and discover the facilities that should be available. Compare the two monitor programs available from Motorola and Intel.

8 Examine the instruction sets of the Motorola and Intel microprocessors in respect of branching, transfer of control, interrupts. Use the instruction sets to produce code segments using different addressing modes.

Index